Parks at sites of castles

Park	Prefecture
Goryoukaku Park	Hokkaido
Matsumae Park	Hokkaido
Hirosaki Park	Aomori
Shiroyama Park	Aomori
Iwate Park	Iwate
Aobayama Park	Miyagi
Masuoka Park	Miyagi
Takashimizu Park	Akita
Senshu Park	Akita
Yokote Park	Akita
Honjo Park	Akita
Kajo Park	Yamagata
Tsuruoka Park	Yamagata
Tsurugajo Park	Fukushima
Shiroyama Park	Fukushima
Baryo Park	Fukushima
Kasumigajo Park	Fukushima
Utsunomiya castle ruins Park	Tochigi
Numata Park	Gunma
Hachigatajo Park	Saitama
Ruins of Sakura Castle Park	Chiba
Shiroyama Park	Chiba
Odawara Castle Ruins of Park	Kanagawa
Ishigakiyama Ichiyajo Historical Park	Kanagawa
Shibata Castle Ruins of Park	Niigata
Takada Park	Niigata
Takaoka Kojyo Park	Toyama
Kanazawa Castle Park	Ishikawa
Kameyama Park	Fukui
Kasumigajo Castle Park	Fukui
Kitanosho Castle Park	Fukui
Maizuru Castle Park	Yamanashi
Matsushiro Castle Park	Nagano
Matsumoto Castle Park	Nagano
Uedajo Castle Park	Nagano
Komoro Castle Ruins · Kaikoen Park	Nagano

Park	Prefecture
Takato Castle Ruins Park	Nagano
Ogaki Park	Gifu
Shiroyama Park	Gifu
Gifu Park	Gifu
Sumpu Castle Park	Shizuoka
Meijo Park	Aichi
Toyohashi Park	Aichi
Okazaki Park	Aichi
Matsuzaka Park	Mie
Kyuka Park	Mie
Zeze Castle Ruins Park	Shiga
Ho Park	Shiga
Fukuchiyama Castle Park	Kyoto
Shoryu-ji Castle Park	Kyoto
Osaka Castle Park	Osaka
Himeji Park	Hyogo
Akashi Park	Hyogo
Ako Castle Park	Hyogo
Shiroyama Park	Shimane
Takamatsu Castle Park	Okayama
Ujo Castle Park	Okayama
Kakuzan Park	Okayama
Chuo Park	Hiroshima
Shizuki Park	Yamaguchi
Tokushima Central Park	Tokushima
Tamamo Park	Kagawa
Kameyama Park	Kagawa
Dogo Kouen Park	Ehime
Shiroyama Park (Matsuyama Castle)	Ehime
Fukiage Park	Ehime
Shiroyama Park (Uwajima Castle)	Ehime
Shiroyama Park (Ozu Castle)	Ehime
Kochi Park	Kochi
Tamematsu Park	Kochi
Katsuyama Park	Fukuoka
Maizuru Park	Fukuoka

Park	Prefecture
Saga Castle Park	Saga
Asahigaoka Park	Saga
Shimabara Castle Park	Nagasaki
Joyama Park	Nagasaki
Omura Park	Nagasaki
Kumamoto Castle Park	Kumamoto
Historical Park Kikuchi Castle	Kumamoto
Usuki Park	Oita
Oka Castle Ruins	Oita
Shiroyama Park	Miyazaki
Shurijo Castle Park	Okinawa
Urasoe Big Park	Okinawa
Zakimi Castle Ruin Park	Okinawa

Hirosaki Park, Aomori Prefecture

Himeji Pa...

Is there any of these parks near your home?

01303

Daiki

Nanami

2

Let's learn mathematics together!

Yui

Hiroto

Important words and rules

Rules that you found

Let's deepen.

You will want to learn much more.

Want to connect

Solve new problems.

3

 # How many people?

Problem: How can we represent numbers larger than 10000?

10 Large Numbers
Let's explore how to represent numbers and their structure.

1 Numbers greater than the thousands place

Want to know Numbers larger than 10000

1 How many admission tickets were sold in total at the baseball game?

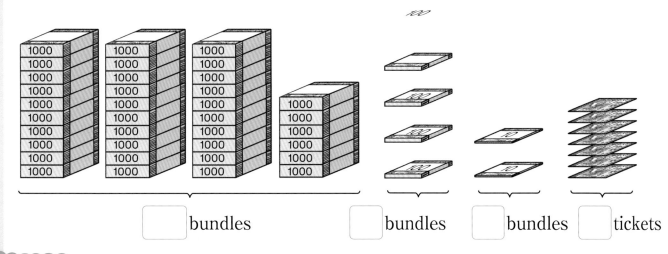

☐ bundles   ☐ bundles   ☐ bundles   ☐ tickets

Want to explore

① How many bundles of 1000, 100, and 10 tickets are there respectively? How many admission tickets are not in bundles?

The number that gathers 6 sets of 1000 is six thousand.
Yui

10 sets of 1000 make 10000, so...

Daiki

▶ Purpose How can we represent numbers larger than 10000?

② How many bundles of 10000 tickets can you make?

The number that gathers 3 sets of ten thousand (万) is written as 30000 and read as **thirty thousand** (三万). Also written as **30 thousand** (3万).

Want to represent

③ How many admission tickets are there altogether?

Way to see and think

Each place value is represented by how many units there are when each bundle is considered as one unit.

	bundles of ten thousand		bundles of one thousand		bundles of one hundred		bundles of ten		tickets

Ten thousands place	Thousands place	Hundreds place	Tens place	Ones place

Summary

The number that is the sum of 3 sets of ten thousand, 6 sets of one thousand, 4 sets of one hundred, 2 sets of ten, and 7 ones is written as 36427 and read as **thirty-six thousand four hundred twenty-seven**.

The next place higher than the thousands place is the **ten thousands place**.

Want to confirm

1 Let's write the following numbers in numerals and read them carefully.

① The number that is the sum of 2 sets of ten thousand, 4 sets of one thousand, 9 sets of one hundred, 1 set of ten, and 8 ones.

② The number that is the sum of 7 sets of ten thousand and 860.

③ The number that is the sum of 8 sets of ten thousand and 9 sets of ten.

④ The number that gathers 4 sets of ten thousand.

2 Let's write the following numbers in numerals.

① eighty-six thousand two hundred fifty-nine

② fifty thousand thirty-two

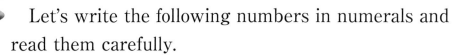

Ten thousands place	Thousands place	Hundreds place	Tens place	Ones place
①				
②				
③				
④				

Want to connect

Are there place values larger than the ten thousands place?

Hiroto

2 The table on the right shows the number of visitors at a theme park in one week, one month, and one year. Let's read each number of visitors.

I week	286837 visitors
I month	1178690 visitors
I year	14602500 visitors

Nanami: It becomes a large number with a place value higher than the ten thousands place.

Daiki: The same as we did until now, the place value moves when we gather 10 sets of ten thousand?

♈Purpose How can we represent a place value higher than the ten thousands place?

10 sets of one thousand is ten thousand →

10 sets of ten thousand is one hundred thousand →

10 sets of one hundred thousand is one million →

10 sets of one million is ten million →

			1	0	0	0	0
		1	0	0	0	0	0
	1	0	0	0	0	0	0
1	0	0	0	0	0	0	0

Way to see and think

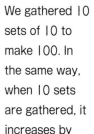

We gathered 10 sets of 10 to make 100. In the same way, when 10 sets are gathered, it increases by one 0.

Want to represent

① Let's write each number of visitors.

	Ten millions place	Millions place	Hundred thousands place	Ten thousands place	Thousands place	Hundreds place	Tens place	Ones place	
I week									visitors
I month									visitors
I year									visitors

The ones, tens, and hundreds become repeated.

♟Summary

From the ten thousands place, each higher place is called the **hundred thousands place**, the **millions place**, and the **ten millions place**.

② The number of visitors in one week is read as two hundred eighty-six thousand eight hundred thirty-seven. Let's read the number of visitors in one month and one year.

③ Let's fill in each ☐ with the appropriate numbers.

ⓐ 286837

The sum of ☐ sets of one hundred thousand, ☐ sets of ten thousand, ☐ sets of one thousand, ☐ sets of one hundred, ☐ sets of ten, and ☐ ones.

ⓑ 1178690

The sum of ☐ sets of one million, ☐ sets of one hundred thousand, ☐ sets of ten thousand, ☐ sets of one thousand, ☐ sets of one hundred, ☐ sets of ten.

ⓒ 14602500

The sum of ☐ sets of ten million, ☐ sets of one million, ☐ sets of one hundred thousand, ☐ sets of one thousand, and ☐ sets of one hundred.

Want to confirm

3 Let's read the following numbers.

① The number of elementary, junior, and high school students in Japan in 2017 was 13117227.

② The number of personal computers made in Japan in 2016 was 4810736.

That's it. How to read a number in Japanese.

Counting from the ones place, separate the numbers in the fourth and fifth places and read.

1306 2281

ten thousands (万)

4 Let's write the following numbers in numerals.

① The population of Chiba prefecture in 2017 was six million two hundred fifty-five thousand ninety-one.

② The number of cars in Japan in 2017 was eighty-one million seven hundred eight thousand one hundred sixty-two.

❷ Structure of large numbers

Want to know

1 The population in Tokyo was 13740000 people in 2017. How many sets of 10000 are gathered in this number?

Way to see and think

What should we consider as the unit?

① Let's use the table shown on the right and think how many sets of 10000 are gathered in 13740000.

② How many sets of 1000 are gathered in 13740000?

Ten millions	Millions	Hundred thousands	Ten thousands	Thousands	Hundreds	Tens	Ones
			万				
1	3	7	4	0	0	0	0
				1	0	0	0	0

13740000 is also written as
13 million 740 thousand (1374万).

"Ones place" can simply be written as "Ones."

Want to confirm

 How many sets of 10000 are gathered in 4080000? How many sets of 1000 are gathered in 4080000?

Want to try

 What number gathers 35 sets of 10000?

Words

Ten thousand

【万】

The character 【万】 means "all" or "large number" and is used in terms such as "万事 banji" (everything), "万能 bannō" (almighty), and "万病 manbyō" (all kinds of diseases).

2 What does one scale represent in the following line of number?

Also, what numbers are represented by ① and ② ?

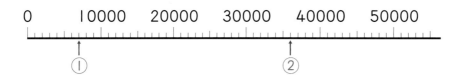

A straight line with marked points that are equally spaced and wherein every point on the line corresponds to a number is called a **number line**.

On the number line, numbers become larger as you move toward the right.

3 Let's fill in each ☐ with the appropriate numbers.

① 1000
0

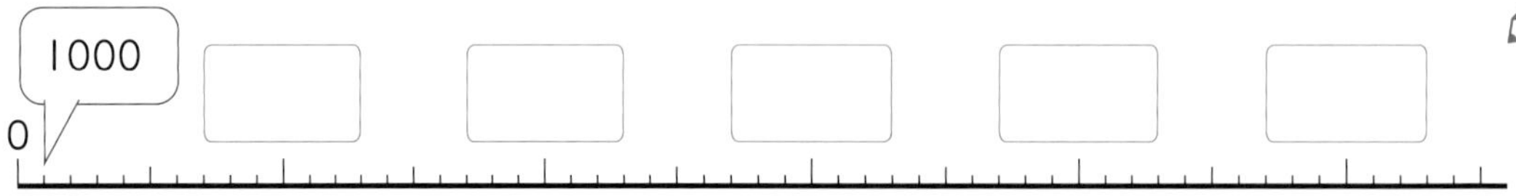

② 10000
0

0   10 million   20 million   30 million   40 million   50 milli

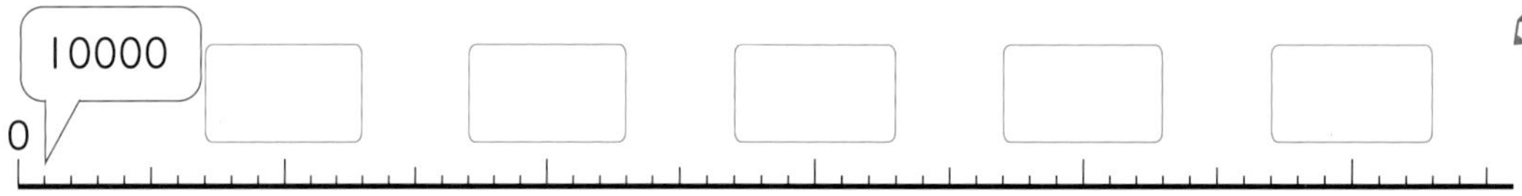

4 Let's explore the following two number lines.

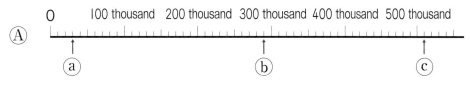

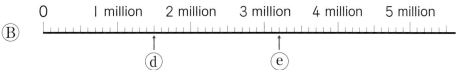

① What does one scale represent in each number line?

② What numbers are represented by ⓐ, ⓑ, ⓒ, ⓓ, and ⓔ?

5 Let's fill in each ☐ with the appropriate numbers.

① ⊣ 99998 ⊢ 99999 ⊢ ☐ ⊣ 100001 ⊢ ☐ ⊢

② ⊣ 2 million 900 thousand ⊢ 2 million 950 thousand ⊢ ☐ ⊣ 3 million 50 thousand ⊢ ☐ ⊢

③ ⊣ 5000 ⊢ ☐ ⊣ 15000 ⊢ ☐ ⊣ 25000 ⊢

6 On the number line below, number Ⓐ gathers how many sets of ten million?

> The number that gathers 10 sets of ten million (千万) is written as 100000000 and read as **one hundred million** (一億). Also written as 100 **million** (1億).

Way to see and think

The place value moves in the same way as when 10 sets of one thousand are gathered.

60 million   70 million   80 million   90 million   Ⓐ

3 Let's write the following numbers in descending order.

386020　1290000　378916

Ten millions	Millions	Hundred thousands	Ten thousands	Thousands	Hundreds	Tens	Ones
			万				

We should compare numbers from the highest place value.

Hiroto

7 Let's represent the relationship between the following two numbers by using > or <.

45000 ☐ 140000

The symbols > and < are called **inequality signs**. These symbols are used to compare two numbers or math expressions, to represent that one is larger or smaller than the other.

8 Let's compare the following pair of numbers and fill in each ☐ with > or <.

① 54300 ☐ 64100　② 17300 ☐ 17030

9 Let's arrange the following numbers in ascending order.

① (30001, 190000, 210003, 99900)

② (400000, 94000, 170000, 240000)

3 10 times, 100 times, 1000 times, and divided by 10

Want to explore 10 times of a number

1 Let's explore the number that gathers 10 sets of 20.

Daiki

2 sets is 2 times, 3 sets is 3 times, so...

"10 sets" and "10 times" have the same meaning.

Yui

Purpose 10 times of a number becomes what kind of number?

① Let's explain based on the diagram shown below.

 → 100

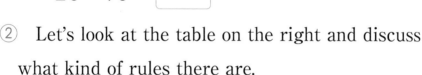 → 100

20 × 10 = [ ]

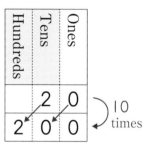

 10 times

② Let's look at the table on the right and discuss what kind of rules there are.

Way to see and think

The idea is to decompose 25 into 20 and 5.

Want to try

 1 What is 10 times of 25?

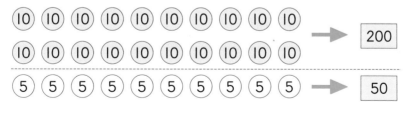 → 200

→ 50

25 × 10 = [ ]

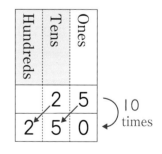 10 times

Summary

When 10 times of a number is done, each digit of that number moves to the next higher place and then 0 is added at the end.

13

2 What is 100 times of 25?

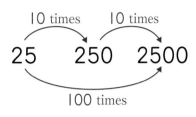

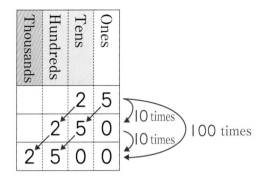

$$25 \times 100 = \boxed{}$$

 3 What is 1000 times of 25?

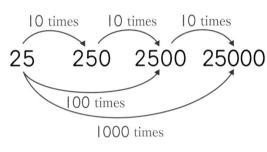

$$25 \times 1000 = \boxed{}$$

When 100 times of a number is done, each digit of that number moves two places up and then 00 is added at the end. Also, when 1000 times of a number is done, each digit of that number moves three places up and then 000 is added at the end.

 4 Let's find the numbers that are 10 times, 100 times, and 1000 times of the following numbers.

① 70   ② 500   ③ 640   ④ 850

 5 What is 100 times of 100? Also, what is 10 times of 10 million?

14

2 Let's explore the number that is 150 divided by 10.

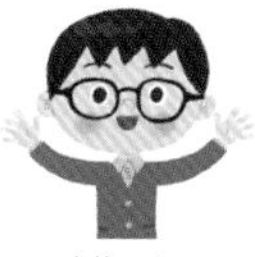

Hiroto: What will happen when dividing by 10?

When 10 times is done the numerals move one place up, so...

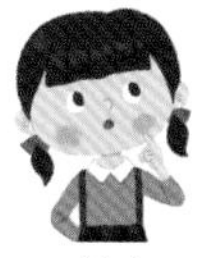

Yui

Purpose A number divided by 10 becomes what kind of number?

Want to explain

① Let's explain based on the diagram shown below.

$$150 \div 10 = \boxed{}$$

② Let's look at the table on the right and discuss what kind of rules there are.

Way to see and think

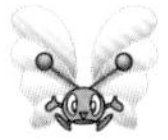

The idea is to decompose 150 into 100 and 50.

Hundreds	Tens	Ones	
1	5	0	divided
	1	5	by 10

Summary

If any number with a 0 in the ones place is divided by 10, each digit of that number moves to the next lower place and the 0 in the ones place disappears.

Want to confirm

6 Let's divide the following numbers by 10.

① 700　　② 5000　　③ 6400　　④ 8500

7 Let's find 10 times of 35. Then, divide the answer by 10. What can we understand from this?

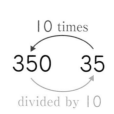

10 times

350　35

divided by 10

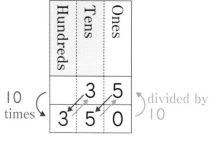

4 Addition and subtraction of large numbers

1

In 2017, the number of children in 3rd grade in all elementary schools in Japan was 1080000 and the number in 4th grade was 1090000. How many children were there in 3rd grade and 4th grade altogether?

① Let's write a math expression.

☐ + ☐

Writing many zeros consumes time.

Hiroto

② Let's think about how to calculate.

Ten millions	Millions	Hundred thousands	Ten thousands	Thousands	Hundreds	Tens	Ones
			万				
1	0	8	0	0	0	0	0
1	0	9	0	0	0	0	0

Way to see and think

You can consider 10 thousands as one unit.

 What is the difference between the number of children in 3rd grade and 4th grade in **1**?

☐ − ☐

Way to see and think

You can consider what is one unit in the same way as in the addition.

 Let's calculate the following.

① 2 million 450 thousand + 270 thousand

② 4 million 680 thousand − 1 million 170 thousand

③ 90 million + 10 million   ④ 98 million − 20 million

⑤ 210000 + 370000   ⑥ 530000 − 180000

Way to see and think

Considering 10 thousand as the unit, the addition and subtraction of sets of ten thousand can be calculated in the same way as done until now.

3 Let's use each card numbered 1, 2, 3, 4, 5, 6, 7, and 8 only one time to make addition and subtraction problems of 4-digit numbers. Then, let's calculate.

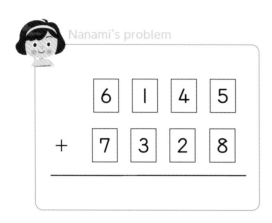

Nanami's problem

```
  6 1 4 5
+ 7 3 2 8
```

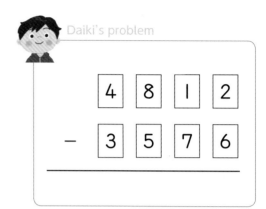

Daiki's problem

```
  4 8 1 2
- 3 5 7 6
```

① Let's make the addition problem that has the largest answer.

Yui: When will the addition have the largest answer?

Hiroto: The two largest numbers in the thousands place give the largest sum.

② Let's make the subtraction problem that has the smallest answer.

Daiki: When will the subtraction have the smallest answer?

Nanami: We should make numbers close to each other.

4 Let's calculate 7653 + 2784.

```
  7 6 5 3
+ 2 7 8 4
```

Way to see and think

The way of calculation is the same even when the number has many digits.

17

What you can do now

□ Understanding how to represent numbers.

1 Let's answer the following problems.

① Let's read the following numbers.

ⓐ 542915   ⓑ 6380000

② Let's write the following numbers in numerals.

ⓐ three million eight hundred sixty-seven thousand one hundred twenty

ⓑ fifty-four million

□ Understanding the structure of large numbers.

2 Let's fill in each □ with the appropriate numbers.

① 480270 is the sum of 48 sets of ☐ and 27 sets of ☐ .

② 25607 is the sum of ☐ sets of ten thousand, ☐ sets of one thousand, ☐ sets of one hundred, and ☐ ones.

③ 59 million 200 thousand is the sum of ☐ sets of ten thousand.

□ Understanding the structure of number lines.

3 What numbers are represented by ①, ②, and ③?

```
        0    1 million  2 million  3 million  4 million  5 million
        |_____|_____|_____|_____|_____|
              ↑               ↑                      ↑
              ①               ②                      ③
```

□ Understanding how to compare numbers.

4 Let's fill in each □ with the appropriate inequality sign.

① 333300 ☐ 34330   ② 5482941 ☐ 5482899

□ Understanding how numbers change by doing 10 times or dividing by 10.

5 Let's find the numbers that are 10 times, 100 times, and 1000 times of the following numbers. Also, let's divide the following numbers by 10.

① 200   ② 1080   ③ 30000   ④ 40500

□ Can add and subtract large numbers.

6 Let's calculate the following.

① 620 thousand + 850 thousand   ② 350000 + 470000

③ 1 million 50 thousand − 870 thousand   ④ 91000 − 74000

Supplementary Problems ········ p.144

Usefulness and efficiency of learning

1 Let's write the number that is 1 smaller than 100 million in numerals. Also, let's read it.

Understanding how to represent numbers.

2 Let's fill in each ☐ with the appropriate numbers.

① 31 million 620 thousand is the sum of ☐ sets of ten million, ☐ sets of one million, ☐ sets of one hundred thousand, and ☐ sets of ten thousand.

Understanding the structure of large numbers.

② 107800 is the sum of ☐ sets of 100000 and 78 sets of ☐ .

③ 69301 is the sum of ☐ sets of ten thousand, ☐ sets of one thousand, ☐ sets of one hundred, and ☐ ones.

④ The numerals on the thousands place, ten thousands place, and hundred thousands place of 503408 are ☐ , ☐ , and ☐ respectively.

3 Let's indicate 180 thousand, 360 thousand, and 520 thousand by using an ↑ on the following number line.

Understanding the structure of number lines.

```
0           200 thousand    400 thousand    600 thousand
|_____|_____|_____|_____|_____|
```

4 Let's arrange the following numbers in ascending order.

① (101000, 98900, 50009, 110000)
② (201000, 90700, 200000, 199900)

Understanding how to compare numbers.

5 Let's find the number that is 100 times of the following numbers and then divide by 10.

① 23   ② 40   ③ 111   ④ 605

Understanding how numbers change by doing 10 times or dividing by 10.

6 The population of East City and West City is 130000 people and 260000 people respectively.

① After adding the population of both cities, how many people are there?

② Which city has a larger population and by how many people?

Can add and subtract large numbers.

Places with an equal distance?

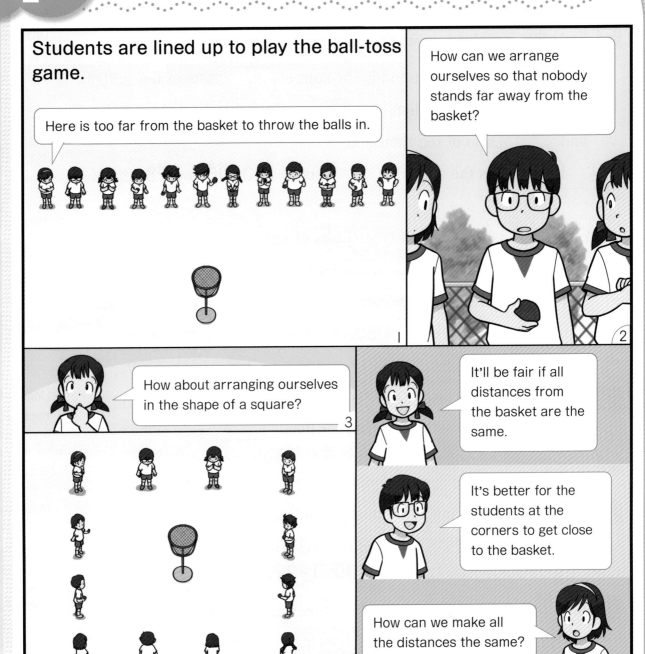

Students are lined up to play the ball-toss game.

How can we arrange ourselves so that nobody stands far away from the basket?

Here is too far from the basket to throw the balls in.

How about arranging ourselves in the shape of a square?

It'll be fair if all distances from the basket are the same.

It's better for the students at the corners to get close to the basket.

How can we make all the distances the same?

Problem   What would the arrangement look like when all the distances to the basket are the same?

11 Circles and Spheres
Let's explore the properties of round shapes and how to draw them.

❶ Circles

Want to think Shape where all the distances are the same

1 How should they arrange themselves to have the same distance from the basket? Let's consider placing small things such as marbles or blocks on the diagram shown below.

Nanami

When they are arranged with the same distance from × at the center, the shape would be...

It would be something round.

Hiroto

⊙ Purpose What kind of shape will it become when they arrange themselves to have the same distance from a point?

Let's draw many points that are **3 cm** away from point A. What kind of shape will it become?

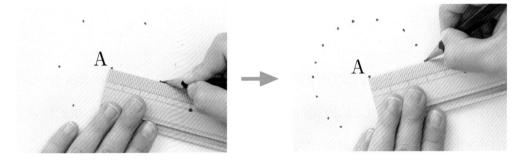

⚘ Summary

Drawing many points at the same distance from a certain point becomes a round shape.

A round shape that consists of points that have the same distance from one point is called a **circle**. This point in the middle is called the **center** of the circle. The straight line from the center to any point on the circle is called the **radius**. In a circle, the length of every radius is the same.

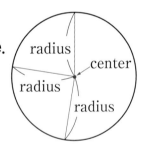

The circle that you have drawn in 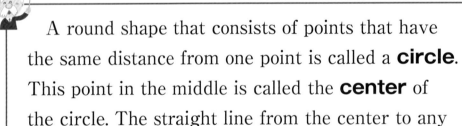 has a radius of **3 cm**. Also, point A is the center of the circle.

 Let's look for circular shapes in our surroundings.

Can we call any round shape a circle such as ⬭?

(Fuji City, Shizuoka Prefecture)

Yui

Want to know How to draw circles

2 As shown below, let's draw circles of various sizes.

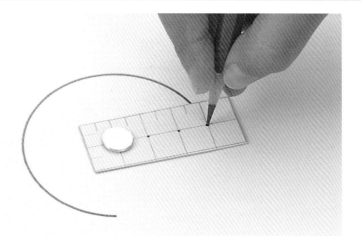

Way to see and think

The circle is drawn by keeping the radius always the same.

Want to confirm

3 Let's draw many radii on the circle shown on the right and explore their length. Also, let's confirm that every radius has the same length.

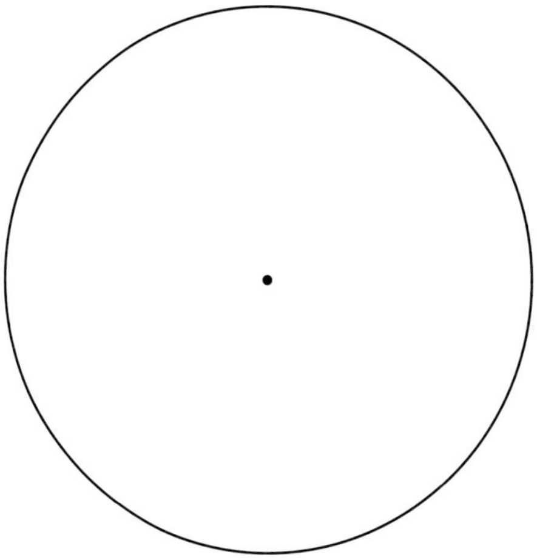

Want to try

4 Let's draw a circle with a radius of 2 m by using a rope at our school ground.

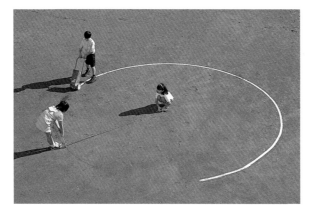

Way to see and think

You can draw circles in the same way when radii are large.

3 A compass is useful to draw circles. A circle with a radius of 1 cm was drawn below. Using the same center, let's draw circles with a radius of 2 cm, 3 cm, 4 cm, and 5 cm.

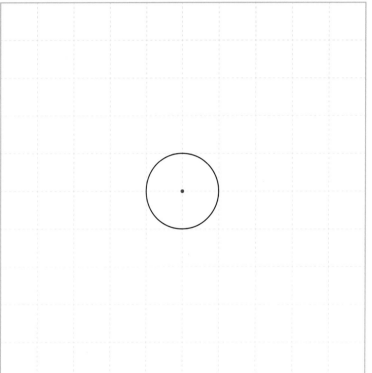

Let's draw circles using a compass and referred to the instructions shown below.

Let's draw circles with various radii in your notebook.

How to draw circles with a compass

(1) Remove any pad from underneath the page.

(2) First, determine the center of the circle, and then mark the points according to the length of the radius.

(3) Open the compass based on the length of the radius.

5 Let's explore the circle with point A as the center.

① Let's draw a circle with a radius of **3 cm.**

② Let's draw any radius of the circle and extend the line to the opposite side on the surrounding circle.

Then, let's explore the length of the extended line and discuss what you understood.

A ·

A straight line drawn from one point on the surrounding circle passing through the center of the circle to another point on the surrounding circle is called a **diameter** of the circle.

The length of the diameter is twice the length of the radius.

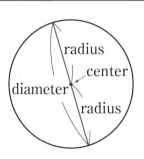

It is good to turn your wrist toward yourself at first for a complete turn without stopping your compass.

(4) Place the compass needle at the center.

(5) Start turning the compass toward your wrist.

(6) Turn the compass without stopping.

01304

Want to know How to draw circles with the same size

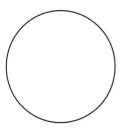

4 Let's draw a circle with the same size of the one shown above by using a compass.

① What should we need to know to draw the circle?

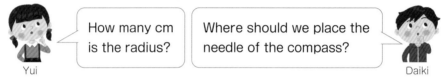

How many cm is the radius?
Where should we place the needle of the compass?
Yui
Daiki

Purpose How can we find the center of a circle?

② Let's draw a circle, cut it out, and fold it as shown below. What can you understand from this?

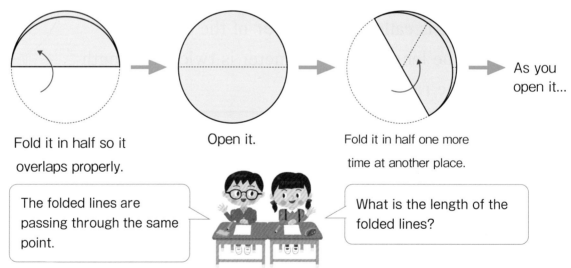

As you open it...
Fold it in half so it overlaps properly.
Open it.
Fold it in half one more time at another place.

The folded lines are passing through the same point.
What is the length of the folded lines?

③ Let's find the center of the circle shown above and draw a circle with the same size.

Summary

After folding two times a circle into halves that overlap properly, the center of the circle can be found at the point where the two folded lines intersect each other.

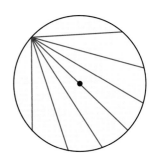

6   Let's draw many straight lines from a point on the surrounding circle to other points on the surrounding circle as shown on the right. What kind of straight line is the longest straight line?

> A diameter is the longest straight line drawn between two points on the surrounding circle. There are many diameters that can be drawn in a circle and their lengths are all the same. All diameters pass through the center of the circle.

Want to improve

Let's find ways to measure the diameter of a circle.

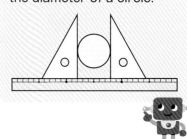

7   How many cm is the diameter of the circle shown on the right?

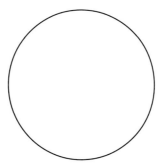

Want to deepen

8   Let's draw a circle that fits exactly in a square with a side of 6 cm.

①   How many cm should the radius of the circle be?

②   Let's find the center of the circle.

③   Let's draw the circle that fits exactly inside the square.

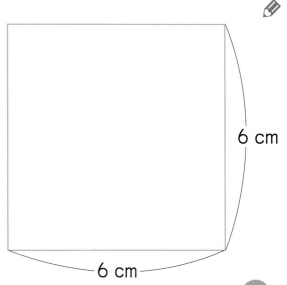

6 cm

6 cm

5 Takuto is going to a park as shown on the map below. Which route is closer, Ⓐ or Ⓑ?

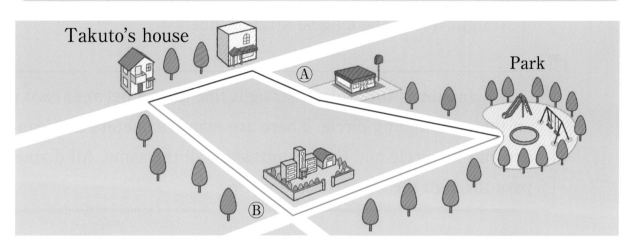

 Daiki: We cannot calculate without the lengths on the map.

We could measure them if the routes were straight lines. Nanami

🌱Purpose How should we compare the length of bent lines?

① Let's transfer the length of the routes Ⓐ and Ⓑ to the following straight lines by using a compass, and explore which one is longer.

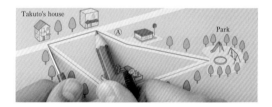

Ⓐ ――――――――――――――――――――――――――――

Ⓑ ――――――――――――――――――――――――――――

😊Summary

We can compare the length of routes by transferring each length to a straight line by using a compass.

9 Let's cut the following straight line into segments of **3 cm** by using a compass.

A compass can be used to cut lines into the same length.

 10 Let's compare the length of Ⓐ, Ⓑ, and Ⓒ by using a compass.

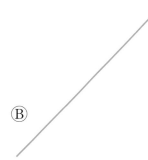

A compass can also be used to compare lengths.

We can use a compass not only for drawing circles but also for transferring lengths, cutting lines, and comparing lengths.

 11 Let's draw beautiful patterns and interesting shapes by using a compass.

①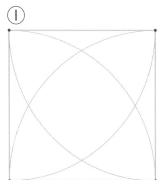

②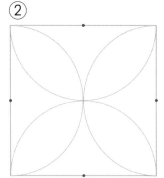

Tomihiro Art Museum
(Midori City, Gunma Prefecture)

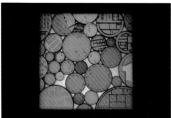

Model of Tomihiro Art Museum

2 Spheres

Want to explore

1 Let's try to explore what kind of shapes can be seen when the following objects are viewed from the side or from the top.

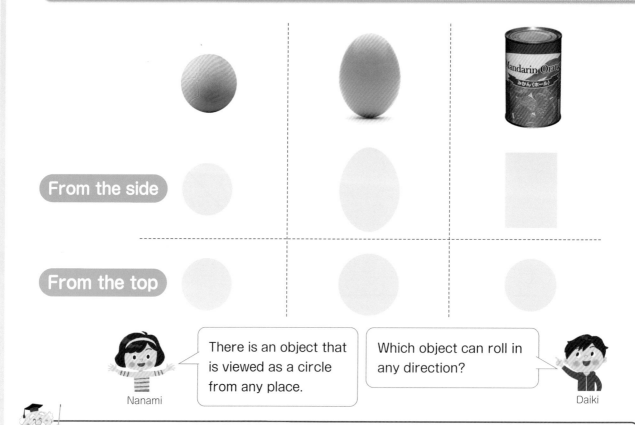

From the side

From the top

There is an object that is viewed as a circle from any place.

Nanami

Which object can roll in any direction?

Daiki

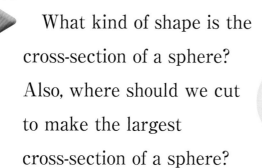

A shape that looks like a circle from any direction is called a **sphere**.

Want to find

1 What kind of shape is the cross-section of a sphere? Also, where should we cut to make the largest cross-section of a sphere?

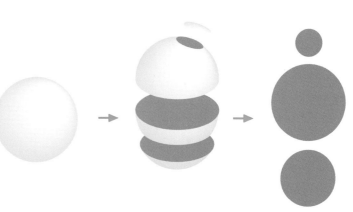

All cross-sections of a sphere are circles. The largest cross-section of a sphere is found when we cut it exactly in half.

When a sphere is cut in half, the center, the radius, and the diameter of the cross-section are called the **center**, **radius**, and **diameter** of the sphere respectively.

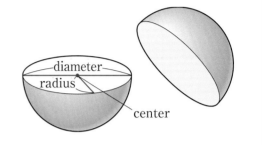

Want to explore

 2 How can we find the diameter of a sphere?

Way to see and think

You also explored the diameter of a circle in the same way.

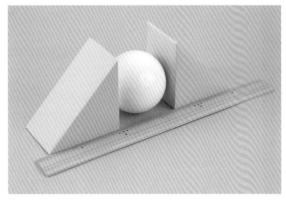

Want to find in our life

 3 Let's look for objects with the shape of a sphere in our surroundings.

Notebook for summarizing

Let's summarize the learning from that day.

Write today's date.

Write the problem.

Let's learn with the purpose.

Organize and write your ideas.

October 20

> Let's think about how to draw round shapes.

Purpose: Let's draw a shape with many points at the same distance from one point.

〈My idea〉
Drawing many points that are 3 cm away from one point.

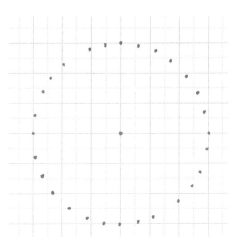

It became the same round shape as when they arranged themselves for the ball-toss game.

Round shape drawn using a grid paper

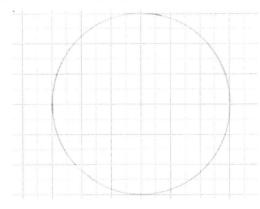

Summary

A round shape that consists of points that have the same distance from one point is called a circle.

    Words: circle
           center
           radius
    There are many radii.

⟨Reflect⟩

· I found that a circle is made of points drawn at the same distance from the center.

· I want to remember the words properly: circle, center, and radius.

Summary

Circle:	A shape that consists of points that have the same distance from one point.
Center:	A point at the middle of the circle.
Radius:	Straight lines drawn from the center to any point on the circle.

Also, write other methods.

Write today's summary using colors and drawing lines around it.

As for reflection, the following must be written:
· what you have found
· what you have noticed
· what you have learned
· what you didn't understand
· what you want to do more.

If you try your best to write a good summary, you will understand better.

What you can do now

☐ Understanding about the diameter and radius of a circle.

1 Let's think about the circle shown on the right.

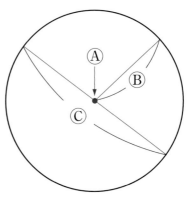

① What is point Ⓐ called?

② What are straight lines Ⓑ and Ⓒ called?

☐ Can draw circles.

2 Let's draw the following circles.

① Circle with a diameter of 4 cm　② Circle with a radius of 4 cm

☐ Understanding how to use a compass.

3 Let's use a compass to compare the length of the following straight lines and arrange symbols Ⓐ, Ⓑ, and Ⓒ from longest to shortest.

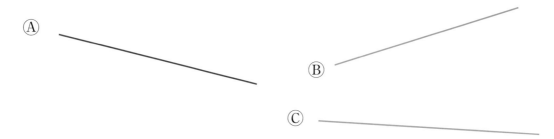

☐ Understanding about spheres.

4 5 balls with a radius of 4 cm were placed in a box as shown on the right.

　Let's find the length, width, and height of the box.

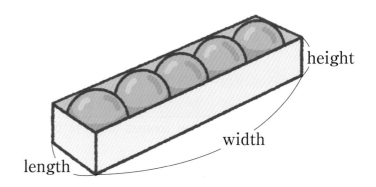

Supplementary Problems
p.145

Usefulness and efficiency of learning

1 Let's fill in each ☐ with the appropriate words.

① The straight line drawn from one point on the surrounding circle passing through the center of the circle to another point on the surrounding circle is called a ☐ .

② The length of the radius of a circle is ☐ of the length of its diameter.

③ A circle has an infinite number of diameters and all have the ☐ length.

④ A ☐ is the longest straight line drawn between two points on the surrounding circle.

☐ Understanding about the diameter and radius of a circle.

2 The figure on the right shows a circle that fits exactly in a square. Let's explore the radius of this circle and draw a circle with the same size.

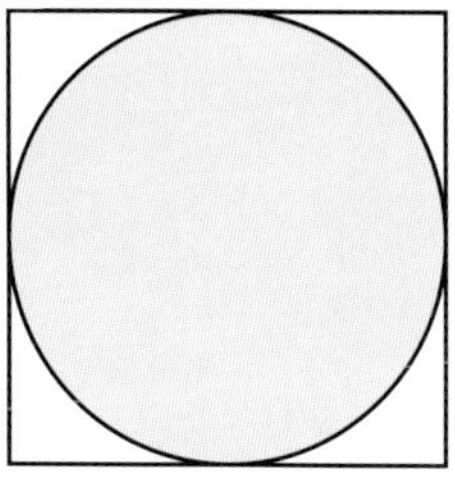

☐ Can draw circles.

3 Which length is longer, the length surrounding the rectangle or that of the square? Let's use a compass and the straight lines drawn below.

Ⓐ　　　　　　　Ⓑ

☐ Understanding how to use a compass.

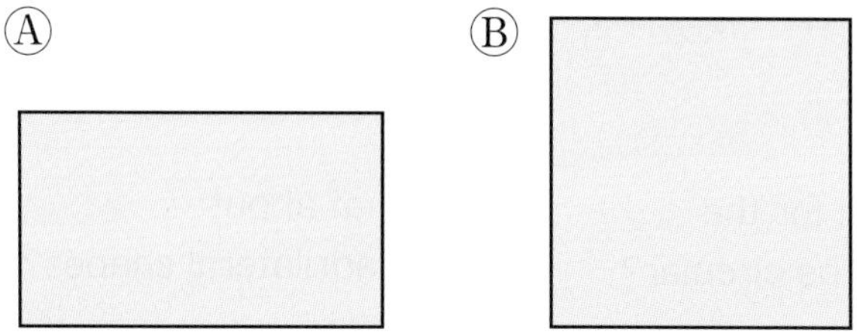

Ⓐ ————————————————————————

Ⓑ ————————————————————————

4 As shown below, 15 balls were placed without overlapping in a box with a length of 18 cm, a width of 30 cm, and a height of 6 cm. How many cm is the diameter of each ball?

☐ Understanding about spheres.

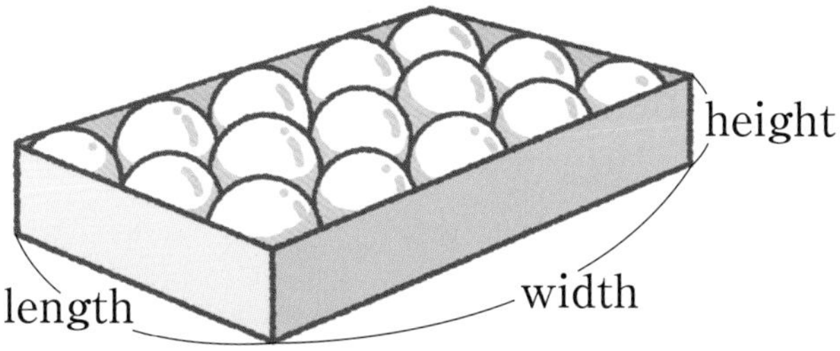

Let's deepen.

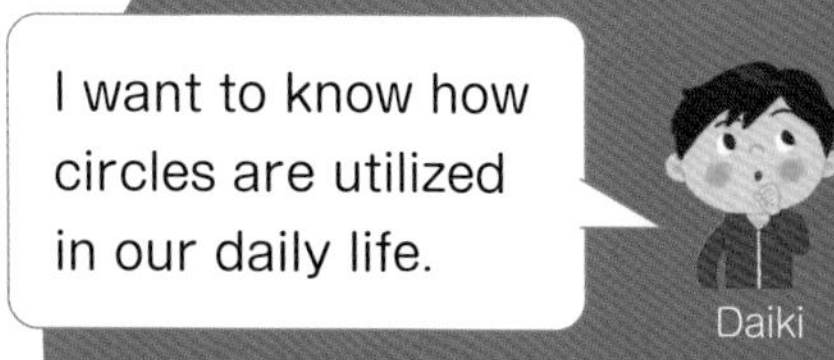

height

length　width

I want to know how circles are utilized in our daily life.

Daiki

35

Deepen.

Why are manhole covers circular?

Want to know

We see many manhole covers on the roads. Some manhole covers have a quadrilateral shape, but most of them have a round shape. Let's think about the reasons why manhole covers are circular.

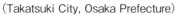
(Takatsuki City, Osaka Prefecture)

(Kobe City, Hyogo Prefecture)

Yui: Is it useful for the covers to be circular?

Hiroto: What about quadrilateral shapes?

① Let's discuss what happens when the covers are removed from the manholes.

Yui: The longest line within a circle is the diameter, so...

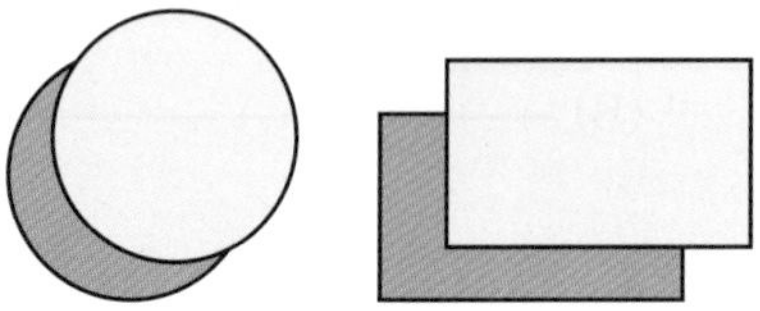

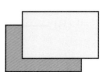
Hiroto: The longest line within a quadrilateral is...

② Let's investigate about other shapes of manhole covers in our country.

Whose top will spin the longest?

Problem How can we compare whose paper-spinning top will spin the longest time?

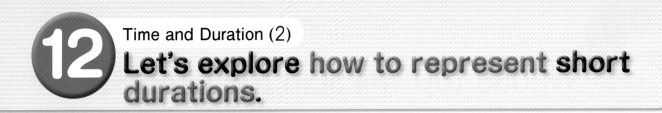

12 Time and Duration (2)
Let's explore how to represent short durations.

1 Let's spin our own paper-spinning tops and investigate whose top will spin the longest time.

Can we compare the spinning duration?

Hiroto

But it may be shorter than 1 minute.

Nanami

Purpose How can we represent durations that are shorter than 1 minute?

A **second** is a unit of time that is shorter than 1 minute. The word second can be written as **sec**.

| 1 minute = 60 seconds |

The second hand takes one second to move from one mark to the next one. The second hand makes one rotation in 60 seconds.

Summary

We can use seconds to represent durations that are shorter than 1 minute.

① Stopwatches are useful to measure short durations. Let's use a stopwatch to measure the duration of a top's spinning time.

2 The table on the right shows the duration of the tops' spinning times. Whose top took the longest time?

Mirei	58 sec
Yugo	1 min 40 sec
Aina	1 min 28 sec
Fumito	104 sec

① Let's represent the durations in seconds.

Yugo 1 min 40 sec = ☐ sec

Aina 1 min 28 sec = ☐ sec

```
  40
+ 60 … 1 min
```

② Let's represent the duration in minutes and seconds.

Fumito 104 sec = ☐ min ☐ sec

```
 104
-  60 … 1 min
```

1 Let's play the game "Exactly 10 seconds" using a stopwatch. Let's close the eyes and try to stop the stopwatch when you think 10 seconds have passed.

I want to try with 30 seconds and 1 minute.

2 Let's find situations that use seconds in our surroundings.

Daiki

What duration does "0:51" represent on the picture shown on the left?

What you can do now

☐ Understanding the relationship between minutes and seconds.

1 Let's fill in each ☐ with the appropriate numbers.

① 1 min = ☐ sec

② 1 min 20 sec = ☐ sec

③ 180 sec = ☐ min

☐ Understanding the size of units for duration.

2 Let's arrange the following durations in descending order.

(150 sec, 1 min 45 sec, 75 sec, 3 min)

Supplementary Problems
•••••••• ➤ p.146

Usefulness and efficiency of learning

1 Let's fill in each ☐ with the appropriate numbers.

① 2 min 10 sec = ☐ sec

② 100 sec = ☐ min ☐ sec

☐ Understanding the relationship between minutes and seconds.

2 Kota, Chisato, and Yuki folded paper planes and tossed them. Kota's plane flew for 1 minute 23 seconds, Chisato's for 46 seconds, and Yuki's for 72 seconds.

Whose plane flew the longest time?

☐ Understanding the size of units for duration.

Amount of water less than 1dL?

Various containers are filled with water.

I've measured the amount of water in the tea cup using 1-dL measuring cups.

Exactly 2 cups, so there are 2 dL.

1 dL   1 dL

1

Let's also measure the amount of water in my mug.

It's 2 dL and a remaining part.

1 dL   1 dL   remaining part

2

My bowl also has a remaining part.

How should we measure remaining parts that are less than 1 dL?

1 dL   1 dL   remaining part

3

Problem   How many dL is the remaining part and the total amount of water?

Let's explore how to represent the size of the remaining part and its structure.

1 How to represent the remaining part

Want to know Amount of water in a mug

1 The following diagram shows the result after measuring the amount of water in the mug by using 1-dL measuring cups. How many dL is the amount of water?

| 1 dL | 1 dL | remaining part |

 Yui: 2 dL and a remaining part which is more than half.

Daiki: How should we measure the remaining part?

Purpose How can we represent the remaining amount of water using dL?

Want to represent

① Let's develop a smaller unit scale by dividing a 1-dL measuring cup into 10 equal parts.

 Dividing things into equal sizes is "equally dividing."

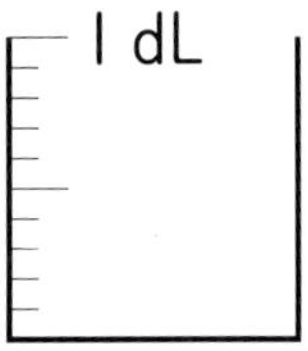

 1 dL

Way to see and think

Should we consider it in the same way as when 1 dL was defined after dividing 1 L into 10 equal parts?

② How can we represent the amount of water in these cups using dL?

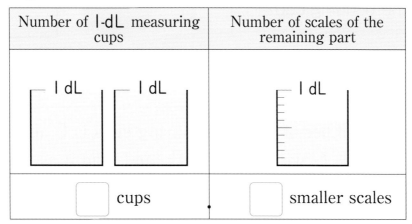

Number of 1-dL measuring cups	Number of scales of the remaining part
☐ cups	☐ smaller scales

We can't say 26 dL.

To distinguish 2 dL from the remaining part, "." is placed in between. This is written as 2.6 dL and read as **"two point six deciliters."**

 1 How many dL of water are there in the following containers?

① Rice bowl

☐ . ☐ dL

② Soup bowl

☐ . ☐ dL

2 How many dL of water are there in the pudding cup?

○ **Purpose** How can we represent a smaller unit scale that is less than I dL ?

☐ . ☐ dL

When the number of I-dL measuring cups is 0 and the number of small scales is 6, we write 0.6 dL and read it as "**zero point six deciliters**."

2 How many dL of water are there in the milk container?

☐ . ☐ dL

○ **Summary**

One small unit scale is 0.1 dL. 0.1 dL is I of the 10 equal parts of I dL.

2.6 dL is 2 dL and 0.6 dL altogether.

Numbers like 2.6, 0.6, and 0.1 are called **decimal numbers** and "." is called the **decimal point**. The place value to the right of the decimal point is called the **tenths place**.

Numbers like 0, 1, 6, and 230 are called **whole numbers**.

2 . 6
···Tenths place
···Decimal point
···Ones place

3 How many dL are the following amounts of water?

① 9 sets of 0.1 dL.   ② 3 dL and 0.5 dL altogether.

Way to see and think

Can we consider it in the same way as with 1-dL cups?

3

The amount of water in a bucket was measured by using 1-L measuring cups. How can we represent the remaining part in decimal numbers?

remaining part

What kind of scale should we place?

Purpose When measuring the amount of water with 1-L measuring cups, can we also represent it in decimal numbers?

① When the 1-L cup is divided into 10 equal parts, the remaining part has 8 sets of the smaller scale. How many L are there altogether?

□ . □ L

Summary

The remaining part that is measured with a 1-L measuring cup can also be represented in decimal numbers. We can make smaller unit scales of 0.1 L by dividing a 1-L measuring cup into 10 equal parts. | 0.1 L = 1 dL |

4 How many cm do ①, ②, and ③ represent in the following number line?

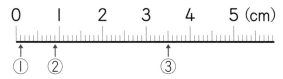

① □ cm

② □ cm

③ □ cm

Way to see and think

We can consider length in the same way as with the amount of water.

5 How many m do ①, ②, and ③ represent in the following number line?

① □ m

② □ m

③ □ m

45

1 The amount of water in the vase is 2.4 dL. This corresponds to how many sets of 0.1 dL?

Daiki: 0.1 dL is one set of 1 dL divided into 10 equal parts, so...

Yui: How did we consider whole numbers?

3 dL

2 dL

1 dL

Purpose Can we represent by considering 0.1 as one unit?

① Besides 2 dL, how many dL is the remaining part?

② Let's color the scale on the right to show the amount of water in the vase.

③ How many sets of 0.1 dL are equal to 2.4 dL?

Summary

We can represent decimal numbers by how many sets of 0.1 there are, such as 2.4 dL is 24 sets of 0.1 dL .

1 How many dL do ①, ②, ③, and ④ represent in the following number line? Also, this represents how many sets of 0.1 dL?

```
0 ┌0.1        1              2              3            4(dL)
  └┬──────────┬──────────────┬──────────────┬────────────
   ↑          ↑              ↑              ↑
   ①          ②              ③              ④
```

2 Let's fill in each ☐ with the appropriate numbers.

① 1.6 dL is ☐ sets of 0.1 dL.   ② 21 sets of 0.1 dL is ☐ dL.

③ 2 sets of 1 dL and 3 sets of 0.1 dL make ☐ dL.

2 Let's explore the following number line.

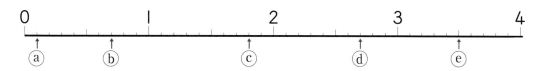

① Let's write the decimal numbers that are pointed on the number line with an ↑.

② How many sets of 0.1 are there in each decimal number ⓐ~ⓔ?

③ Which is larger, 2.1 or 1.9? Let's place an ↓ on the number line to compare both decimal numbers.

④ Which is larger, 0 or 0.1?

⑤ What is 10 sets of 0.1?

> On the number line, numbers become larger as you move toward the right.

3 Let's write the decimal numbers that are pointed on the number line with an ↑.

4 Let's fill in each ☐ with the appropriate inequality sign.

① 3 ☐ 3.1   ② 4.6 ☐ 3.8   ③ 1.2 ☐ 0.9

5 Let's fill in each ☐ with the appropriate numbers.

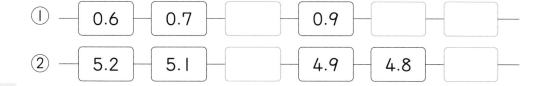

① 0.6 — 0.7 — ☐ — 0.9 — ☐ — ☐

② 5.2 — 5.1 — ☐ — 4.9 — 4.8 — ☐

Want to solve Addition of decimal numbers

1 Ayano's family drank 0.4 L of milk in the morning and 0.5 L of milk in the afternoon. How many L of milk did they drink altogether?

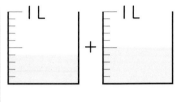

 How can we calculate using decimal numbers?

When finding how much is altogether, we can...

Daiki

Yui

Purpose How should we consider the addition of decimal numbers?

① Let's write a math expression.

② Let's think about how many sets of 0.1 there are.

Summary

When thinking about how many sets of 0.1 there are, the addition of decimal numbers can be calculated in the same way as with whole numbers.

Want to confirm

1 A 0.9 m tape and a 0.3 m tape were joined together. How many m is the total length of the tape?

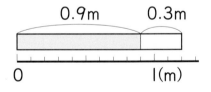

① Let's write a math expression.

② Let's think about how many sets of 0.1 there are.

2 Let's calculate the following.

① 0.2 + 0.5　② 0.8 + 0.1　③ 0.4 + 0.8　④ 0.6 + 0.4

2 There are 2.5 dL of juice in a large cup and 1.3 dL of juice in a small cup. How many dL of juice are there altogether?

① Let's write a math expression.

Way to see and think

We can consider 0.1 as one unit and calculate 25 + 13.

② We can add decimal numbers in vertical form in the same way as in the addition of whole numbers. Let's explain how to calculate in vertical form.

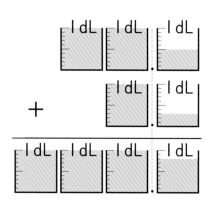

Addition algorithm in vertical form

(1) Align the digits of the numbers according to their places.
(2) Calculate in the same way as with whole numbers.
(3) Write the decimal point for the answer aligned with the above decimal point.

	Ones	Tenths
	2	5
+	1	3
	3	8

 3 Let's calculate the following in vertical form.

When the number in the last place of the answer is 0, what should we do with it?

① 2.3 + 4.8 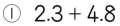　② 0.9 + 7.1 　③ 5 + 3.8

　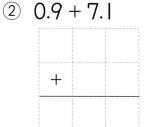　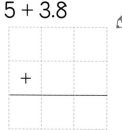

 4 Let's calculate the following in vertical form.

① 3.2 + 1.6　② 3.2 + 1.9　③ 2.9 + 0.3　④ 6 + 3.5

⑤ 4.6 + 7.2　⑥ 2.5 + 7.8　⑦ 2.9 + 8　⑧ 3.8 + 6.2

3 You drank 0.2 L of milk from 0.6 L. How many L of milk are left?

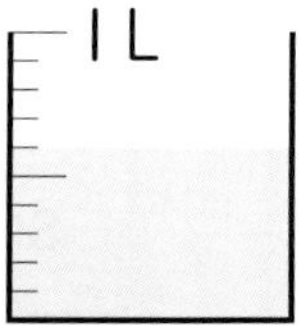

 Nanami — Can we calculate in the same way as in the addition?

Can we also subtract in vertical form? — Hiroto

Purpose Can we subtract decimal numbers in the same way as in the addition?

① Let's write a math expression.

② Let's calculate in vertical form.

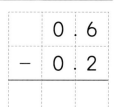

Way to see and think

We should align the digits of the numbers in the same way as in the addition.

Summary

The subtraction of decimal numbers can also be calculated in vertical form in the same way as in the addition.

5 There is a 1.5 m tape. When 0.5 m of this tape is used, how many m are left?

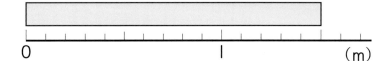

0   1   (m)

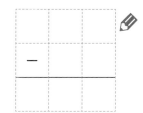

4 Manami has a 1.9 m ribbon and her sister has a 3.5 m ribbon. How many m is the difference between both ribbons?

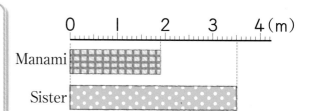

Manami

Sister

① Let's write a math expression.

② Let's think about how to calculate.

③ Let's calculate in vertical form.

```
    3 . 5
 −  1 . 9
```

Way to see and think

Let's think of borrowing in the same way as in the subtraction of whole numbers in vertical form.

6 Let's think about how to calculate the following in vertical form.

① 4.2 − 3.8   ② 4 − 1.8   ③ 1 − 0.7

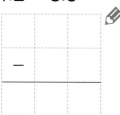 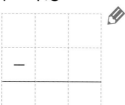 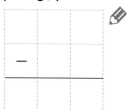

 What happens to the ones place of each answer?

Daiki

We should think of 4 as 4.0.

Yui

7 Let's calculate the following in vertical form.

① 0.7 − 0.3   ② 0.9 − 0.6   ③ 1.3 − 0.6   ④ 1.5 − 0.9

⑤ 3.9 − 1.5   ⑥ 4.1 − 1.7   ⑦ 5.3 − 2.8   ⑧ 4.2 − 1.2

⑨ 3 − 1.2   ⑩ 4 − 0.3   ⑪ 5.2 − 4   ⑫ 8.7 − 8

What you can do now

☐ Understanding the structure of decimal numbers.

1 Let's fill in each ☐ with the appropriate numbers.

① 2.5 is the sum of 2 and ☐.

② ☐ sets of 0.1 is 1.

③ 3 dL and ☐ dL make 3.4 dL.

④ 2.3 cm is ☐ sets of 0.1 cm.

⑤ 1 L and 0.7 L make ☐ L.

⑥ 27 sets of 0.1 m is ☐ m.

☐ Can represent decimal numbers in the number line.

2 Let's write the decimal numbers that are pointed on the number line with an ↑.

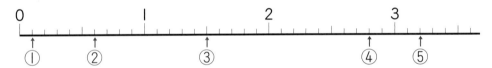

☐ Can compare the size of decimal numbers.

3 Let's fill in each ☐ with the appropriate inequality sign.

① 0.8 ☐ 1.1   ② 2.3 ☐ 3.2   ③ 5.1 ☐ 5

☐ Can add and subtract decimal numbers.

4 Let's calculate the following.

① 3.4 + 1.5   ② 0.6 + 5.2   ③ 0.2 + 0.9

④ 5.7 + 2.6   ⑤ 3.6 + 1.4   ⑥ 8.2 + 1.8

⑦ 5.8 − 3.3   ⑧ 4.6 − 2.7   ⑨ 1.5 − 0.7

⑩ 6.3 − 5.9   ⑪ 7 − 0.7   ⑫ 10 − 2.4

Supplementary Problems p.146

Usefulness and efficiency of learning

1 When the amount of water in a bottle was measured with 1-L measuring cups, it became 1 L and a remaining part. Let's fill in each ☐ with the appropriate numbers.

Understanding the structure of decimal numbers.

① When we want to represent the amount of water by using L as the unit, we need to divide the 1-L cup into ☐ equal parts.

1 L

1 L

remaining part

② The amount of water in the remaining part is ☐ L.

③ The amount of water in the bottle is ☐ L. This amount of water is ☐ sets of 0.1 L.

2 Let's fill in each ☐ with the appropriate numbers.

Understanding the structure of decimal numbers.

① 1.4 is ☐ sets of 0.1.   ② ☐ sets of 0.1 is 2.

③ 4.8 is 4 and ☐ altogether.

3 Let's write the decimal numbers that are pointed on the number line with an ↑.

Can represent decimal numbers in the number line.

```
0
|___|___|___|___|___|___|___|___|___|___|___|___|___|___|
      ↑         ↑              ↑                 ↑   ↑
     0.4       ①              ②                 ③  ④
```

4 Let's arrange the following numbers in descending order.

Can compare the size of decimal numbers.

(1.7, 0.9, 3, 0.1, 10)

5 A small bottle contains 0.8 L of soy sauce and a large bottle contains 1.1 L of soy sauce. How many L are there altogether? Also, how many L is the difference?

Can add and subtract decimal numbers.

Soy Sauce

Soy Sauce

Let's make triangles by using straws with different lengths.

blue

yellow

red

green

A triangle is a shape enclosed by 3 straight lines.

We can use 3 straws to make triangles.

We can make various triangles since there are 4 types of straws with different lengths.

I can see there are some triangles with a similar shape, but can we say they are the same type?

Problem How can we classify the created triangles according to types?

14 Triangles and Angles
Let's explore the properties of triangles and how to draw them.

❶ Isosceles and equilateral triangles

Want to try Classifying triangles according to types

1 The triangles shown below were made by using straws with different lengths. Let's classify the triangles according to types.

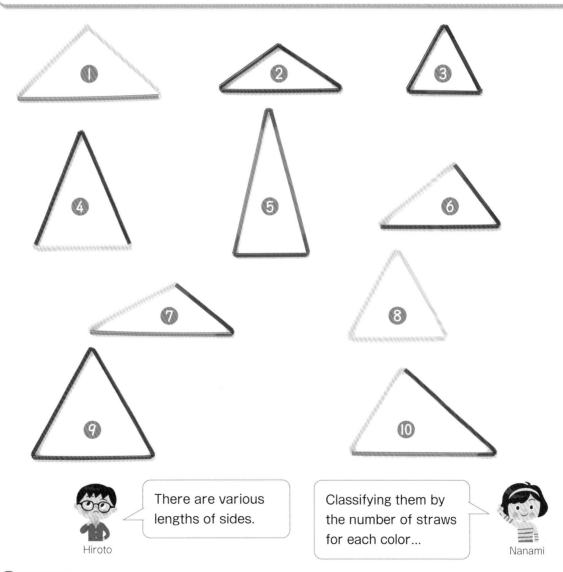

Hiroto: There are various lengths of sides.

Nanami: Classifying them by the number of straws for each color...

▶ Purpose How can we classify triangles according to types?

① Let's explain the classification method followed by Nanami.

Way to see and think

Let's think what kind of straws were used.

Nanami's idea

Ⓐ	Ⓑ	Ⓒ
Blue-Blue-Red 	Blue-Blue-Blue 	Yellow-Blue-Green 
❷	❸	❼

56

② Let's classify the other triangles on page 55 according to Nanami's idea.

Let's cut and paste the triangles on page 158.

③ What kind of triangles were used for the classification type Ⓐ, Ⓑ, and Ⓒ? Let's trace and copy the triangles of each type from the previous page and explore the length of the sides. Let's write the properties on the table shown below.

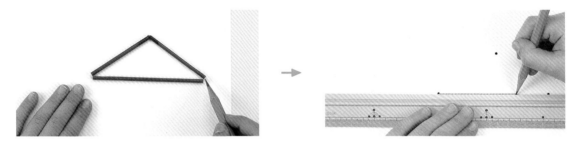

Ⓐ	Ⓑ	Ⓒ

The straws with the same color have the same length.

Hiroto

Summary

We can classify triangles by the number of sides with the same length.

A triangle with two equal sides is called an **isosceles triangle**.

A triangle with three equal sides is called an **equilateral triangle**.

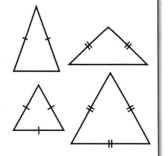

× or ⚹ represent that the lengths of the sides are equal.

Want to classify

2 Let's answer the following questions.

① Which of the following triangles are isosceles triangles?

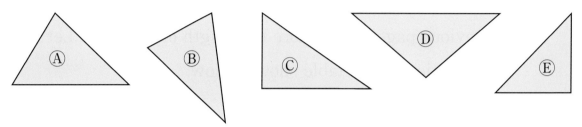

② Which of the following triangles are equilateral triangles?

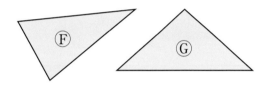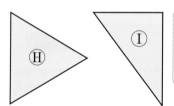

We can use a compass to compare the length of the sides.

Hiroto

Want to find in our life

1 Let's find isosceles and equilateral triangles in our surroundings.

(Kakegawa City, Shizuoka Prefecture)

❷ How to draw triangles

Want to think How to draw isosceles triangles

1 Let's think about how to draw an isosceles triangle whose sides have a length of 3 cm, 4 cm, and 4 cm respectively.

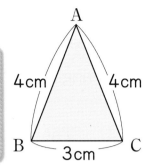

① Side BC has been drawn. Let's try to locate vertex A by watching the diagram shown below.

 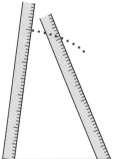

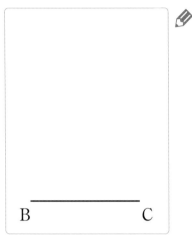

② Let's explain how to draw the triangle by using a compass.

How to draw an isosceles triangle by using a compass

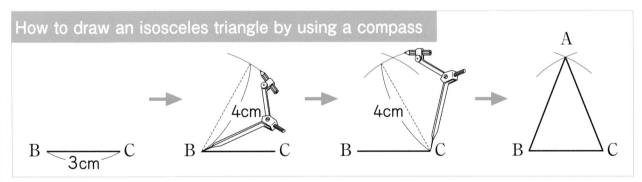

Want to confirm

 Let's draw isosceles triangles whose sides have the following lengths.

① 4 cm, 6 cm, 6 cm   ② 5 cm, 5 cm, 8 cm

Want to deepen

 Can we draw an isosceles triangle whose sides have a length of 3 cm, 3 cm, and 7 cm respectively? In the case we cannot, let's explain the reasons.

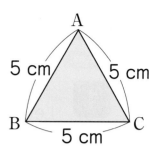

2 Let's think about how to draw an equilateral triangle whose sides have a length of 5 cm.

① The diagram on the right shows the drawing of side BC.

Let's continue drawing the triangle.

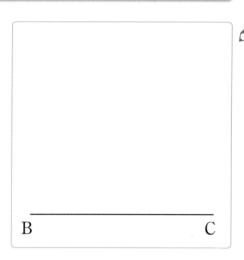

Way to see and think

Can we draw it in the same way as an isosceles triangle?

② How was the equilateral triangle drawn in ①? Let's explain the drawing method.

It is easier to understand if we explain using words like "first", "next", "also", and "lastly."

First, draw the 5-cm side BC .

Next, using a compass, draw a part of a circle with the center at B and a radius of 5 cm.

Also, draw a part of a circle with the center at C and a radius of 5cm in the same way.

Lastly, connect points B and C to the intersection of the parts of the circles that is named point A. This completes the equilateral triangle.

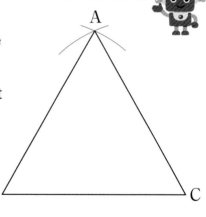

3 Let's draw equilateral triangles whose sides have the following lengths.

① 4 cm   ② 6 cm

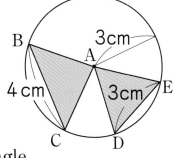

3
Let's draw triangles by using a circle with a radius of 3 cm.

Way to see and think

Sides AB and AC are radii of the circle.

① What kind of triangle is triangle ABC? Let's also explain the reasons.

② Let's draw triangle ABC in the notebook. Also, explain the drawing method.

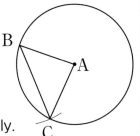

(1) Draw a circle with a radius of 3 cm and name the center A.

(2) Draw a point in the surrounding circle and name it B.

(3) Draw a part of a circle with the center at B and a radius of 4 cm, and name the intersection point C.

(4) Draw straight lines connecting points A, B, and C respectively.

It's easy to understand if we explain the drawing order by placing numbers.

③ What kind of triangle is triangle ADE? Let's also explain the reasons.

④ Let's draw triangle ADE. Continue with the drawing in the diagram shown on the right.

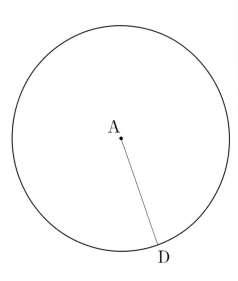

4 Let's draw the following triangles by using a circle.

① An isosceles triangle whose sides have a length of 5 cm, 5 cm, and 4 cm respectively.

② An equilateral triangle whose sides have a length of 4 cm.

Activity

4 Let's make isosceles and equilateral triangles by using origami paper.

① Let's make an isosceles triangle as shown on the right. Also, let's explain the reasons why this becomes an isosceles triangle by paper folding.

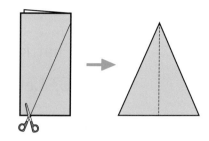

② In the same way, let's make an equilateral triangle.

Nanami

In order to make all sides equal...

It was easy to make an isosceles triangle, but...

Daiki

③ Yui's idea and Hiroto's idea to make equilateral triangles are shown below. Let's compare these ideas.

Way to see and think

Utilize the length of one side of the origami paper.

Yui's idea

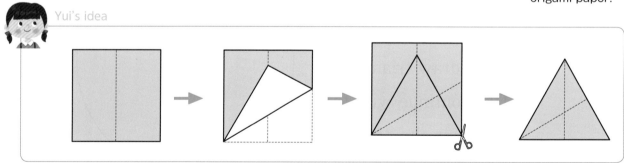

Hiroto's idea

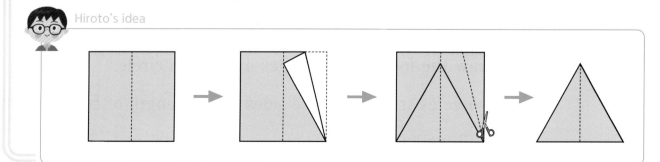

Want to explore The opening between sides

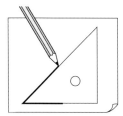
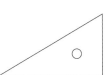

1 Let's trace each corner of the triangle rulers on a paper and explore.

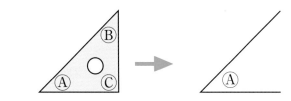
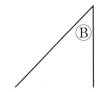

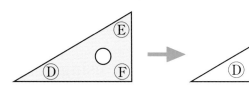

Hiroto

It looks like some of the corners have the same opening and others do not.

It looks like openings beside long sides are wide, but...

Yui

① Which corner is a right angle?

② Which corner is the most pointed?

A figure formed by two straight lines from one point is called an **angle**. The point is called the **vertex** and the two straight lines are called the **sides** of the angle.

The amount of opening between both sides of an angle is called the **size of the angle**.

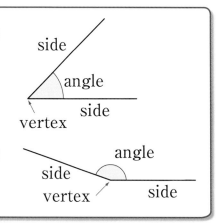

③ Let's compare the size of the angles Ⓐ~Ⓕ and arrange them in descending order.

Want to confirm

1 Let's compare the size of the angles in a large triangle ruler and a small triangle ruler.

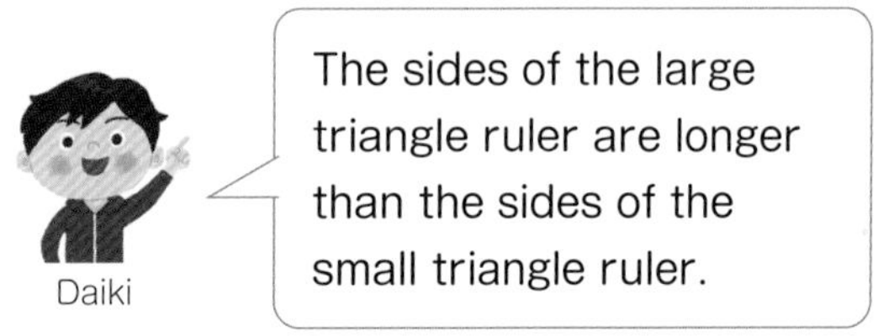

The sides of the large triangle ruler are longer than the sides of the small triangle ruler.
Daiki

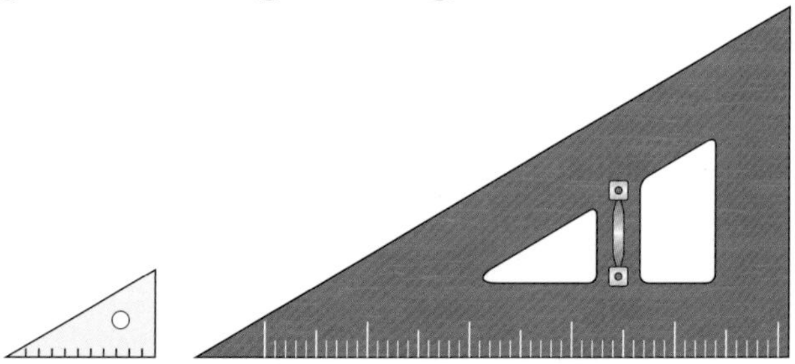

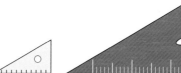

The size of an angle is determined by the amount of opening between both sides and not by the length of the sides.

Want to try

2 Let's compare the size of the angles Ⓖ～Ⓙ by using triangle rulers. Also, let's arrange them in descending order.

Ⓖ
Ⓗ
Ⓘ
Ⓙ

Want to know

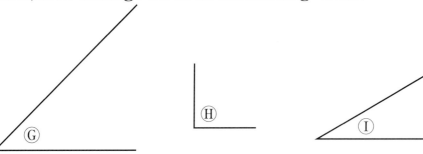

2 What is the relationship between the size of angles in an isosceles triangle? Also, what is the relationship in an equilateral triangle?

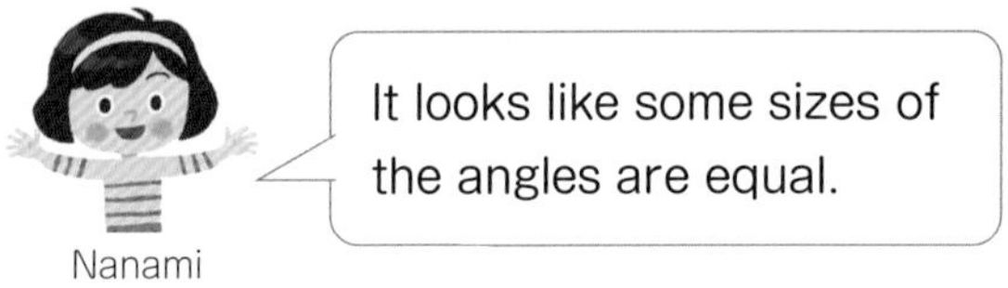

It looks like some sizes of the angles are equal.
Nanami

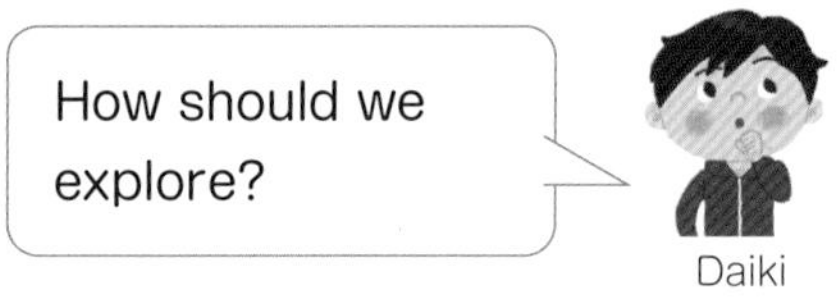

How should we explore?
Daiki

Purpose What is the relationship between the size of angles in an isosceles triangle and equilateral triangle?

① Let's draw an isosceles triangle on a sheet of paper and cut it out. Compare the size of angles Ⓑ and Ⓒ.

② Let's compare the size of angles Ⓐ and Ⓑ.

③ Let's draw an equilateral triangle on a sheet of paper and cut it out. Compare the size of angles Ⓔ and Ⓕ. Also, let's compare the size of angles Ⓓ and Ⓔ.

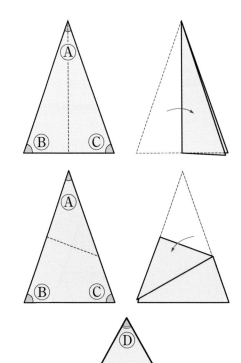

☺ Summary

In an isosceles triangle, the size of two angles are equal.

Also, in an equilateral triangle, the size of the three angles are all equal.

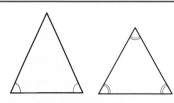
△ or △ shows that the size of the angles are equal.

Want to explain

3 ▶ Let's explore the size of the angles of the triangle ruler shown on the right.

Also, let's explain what you understand.

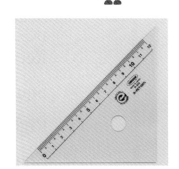

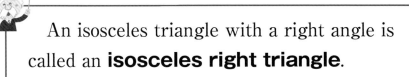
An isosceles triangle with a right angle is called an **isosceles right triangle**.

3 Let's make various shapes by using triangles of the same size.

① Let's try to make shapes by using the isosceles triangles on page 157.

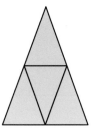 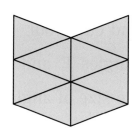 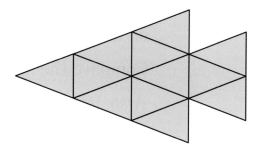

② Let's try to make shapes by using the equilateral triangles on page 157.

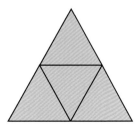

We can make a larger equilateral triangle using the same equilateral triangles.

Nanami

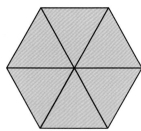

4 Let's explore if we can make the following figures by using two identical triangle rulers.

① Rectangle
② Square
③ Right triangle
④ Equilateral triangle
⑤ Isosceles triangle

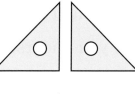

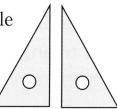

Want to connect

Isosceles triangles and equilateral triangles look similar, is there any relationship between them?

Hiroto

4 What kind of things can we understand as side BC is extended by 1 cm as shown on the following diagrams?

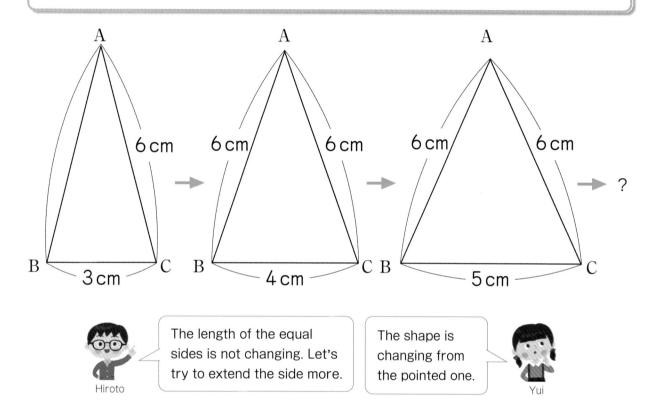

The length of the equal sides is not changing. Let's try to extend the side more.
Hiroto

The shape is changing from the pointed one.
Yui

Purpose What kind of shape does it become when the length of the side of an isosceles triangle changes?

① What kind of triangle does it become when the length of side BC is **6 cm**?

② Let's draw the triangle in ① by using a compass.

③ Let's explore the size of the angles.

Summary

When all the sides of an isosceles triangle are equal, the triangle is an equilateral triangle.

What you can do now

☐ Understanding about the length of sides and the size of angles of isosceles and equilateral triangles.

1 Let's fill in each ☐ with the appropriate numbers.

① An isosceles triangle has ☐ sides with the same length and ☐ angles with the same size.

② An equilateral triangle has ☐ sides with the same length and ☐ angles with the same size.

☐ Can draw isosceles and equilateral triangles.

2 Let's draw the following triangles.

① An isosceles triangle whose sides have a length of 7 cm, 5 cm, and 5 cm.

② An equilateral triangle whose sides have a length of 7 cm.

☐ Understanding the relationship between circles and triangles.

3 The radius of the circle shown on the right is 5 cm and its center is point A. What kind of triangles are ⓐ and ⓑ?

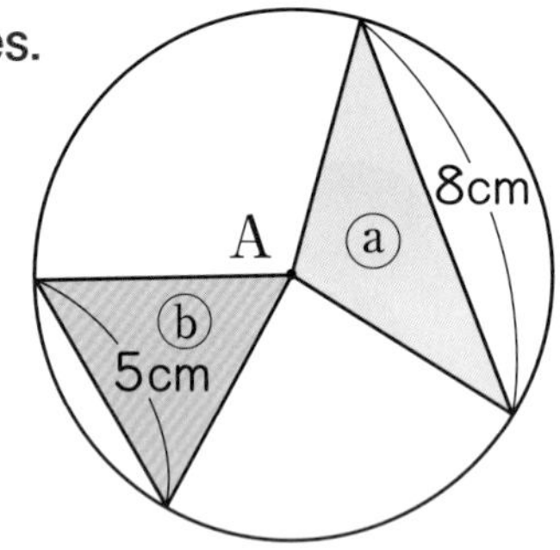

☐ Understanding about angles.

4 Let's fill in each ☐ with the appropriate words.

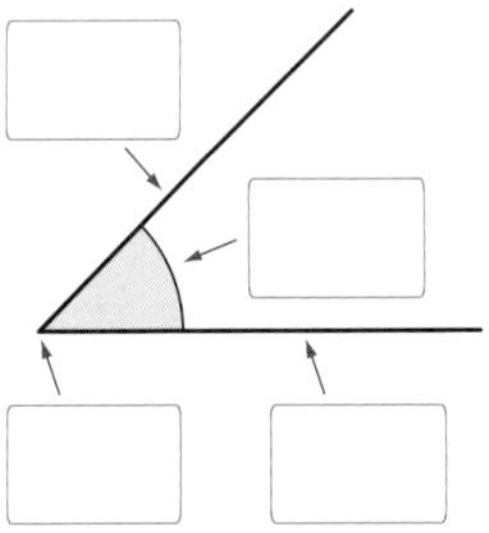

☐ Understanding the relationship between a triangle and the size of the angles.

5 Let's answer about the angles of the triangle rulers shown below.

① Which angle is the smallest?

② Which angle has the same size as angle Ⓐ?

③ Which angle has the same size as angle Ⓕ?

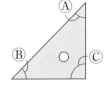
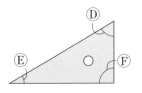

Supplementary Problems ••••••••• p.148

Usefulness and efficiency of learning

1 Which angle has the same size as angle ⓑ in the isosceles triangle on the right?

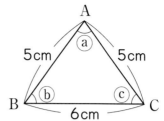

2 Let's answer the following questions about the equilateral triangle shown on the right.

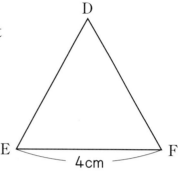

① How many cm is the length of side DE and side DF?

② How many angles have the same size?

3 Let's draw the following triangles. Also, what kind of triangles are these?

① A triangle whose sides have a length of 6 cm, 4 cm, and 4 cm.

② A triangle whose sides have a length of 3 cm.

4 The radii of the two circles shown below have a length of 4 cm and their centers are A and B. BD and AE are the diameters of the circles. Let's draw the same figure and answer the following problems.

① Let's search for isosceles triangles. If you do not know the length of the sides, find a way to measure them.

② Triangle CAB is an equilateral triangle. Let's explain the reasons.

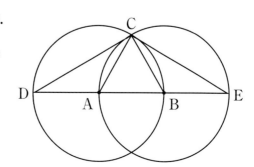

Which route would be better from Shimonoseki Station to Tokyo Station?

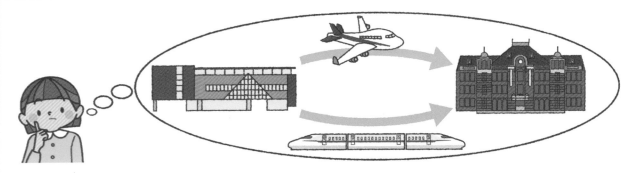

Want to explore

Sakura lives in Shimonoseki city. She will go to the Olympic Stadium in Tokyo with her family in the summer vacation. Since this will be her first time traveling to Tokyo, she asked her sister and brother how to get to Tokyo Station and the cost for one child.

It will be better to go by air in this modern age. Walk 10 minutes to Shimonoseki Station, which is the nearest station from our house. Take a bus to Yamaguchi Ube Airport, that will cost 730 yen with a duration of 1 hour 15 minutes from the station. Wait there for 30 minutes and fly for a duration of 1 hour 35 minutes. The cost is 20690 yen. Finally, it will take you 45 minutes to travel from Haneda Airport to Tokyo Station with a cost of 290 yen.

It will be easy to go by train. It takes 10 minutes to Shimonoseki Station, which is the nearest station from our house. From Shimonoseki Station to Shin-shimonoseki Station it takes an additional 10 minutes. Finally, it will take you 4 hours and 50 minutes to travel from Shin-shimonoseki Station to Tokyo Station by Shinkansen. The cost is 10320 yen.

Think by myself

1 Sakura tried to summarize what they said in a diagram. Let's represent in a diagram each duration and cost for the route provided by her brother and sister. Then, let's write the total duration and cost.

・Route told by her sister

Shimonoseki Station
Ube Airport
10 minutes
1 hour 15 minutes
Tokyo Station
House
0 yen
730 yen

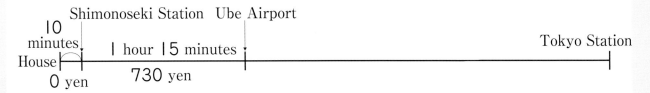

・Route told by her brother

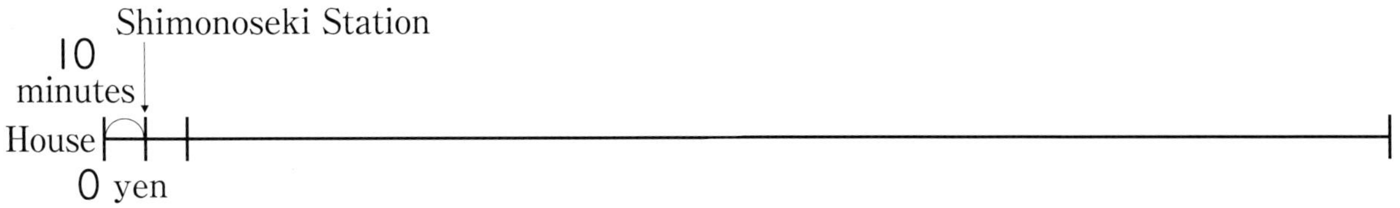

Shimonoseki Station
10 minutes
House
0 yen

Think in groups

2 Sakura calculated the total duration and cost based on the diagram. Even though both ways have good points, she decided to go by Shinkansen. Why did she decide to go by Shinkansen? Let's discuss it in groups by comparing the travel duration and cost.

Think as a class

3 Which route would you choose? Let's explain the reasons to your classmates by using a diagram or something similar.

01305

How many stickers?

There are 4 stickers on each sheet.

Then, how many stickers are there if we buy 8 sheets?

We can find it using the multiplication table.
$4 \times 8 = 32$, so 32 stickers.

Then, how many stickers are there if we buy 10 sheets?

From 4×10, 40 stickers.

What about for 30 sheets?

I think the math expression will be 4×30...

Problem How should we calculate 4×30?

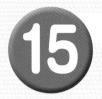

15 Multiplication with 2-digit Numbers

Let's think about how to multiply in vertical form.

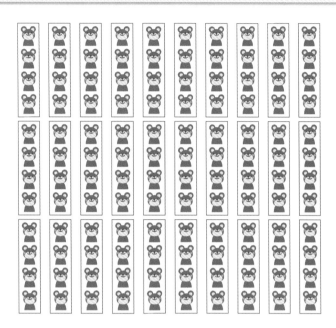

1 Multiplication by tens

Want to solve

Activity

1 There are 4 stickers on each sheet. When there are 30 sheets, how many stickers are there altogether?

① Let's write a math expression to find the total.

☐ × ☐

Number in one unit / How many units

② Let's think about how to calculate.

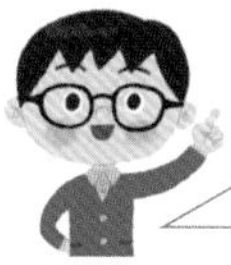

Hiroto

If we divide the 30 sheets into 3 sets of 10 sheets each...

There are 12 stickers on each vertical line and there are 10 sets of them, so...

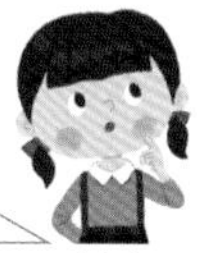

Yui

Way to see and think

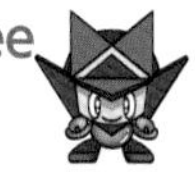

Let's think in the same way as when the multiplicand is sets of ten.

Purpose How should we multiply by tens?

③ Let's explain the ideas of the following children.

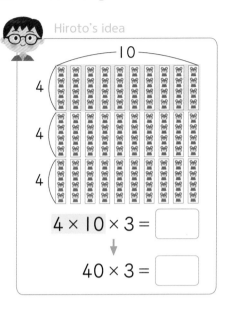

Hiroto's idea

$4 \times 10 \times 3 = \boxed{}$

$40 \times 3 = \boxed{}$

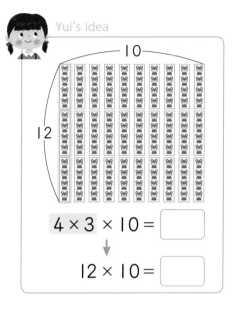

Yui's idea

$4 \times 3 \times 10 = \boxed{}$

$12 \times 10 = \boxed{}$

Way to see and think

Each idea is summarized by horizontal or vertical sets.

Summary

Since 4×30 is 10 times of 4×3, the answer is the same as 4×3 with one 0 at the end.

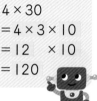

4×30
$= 4 \times 3 \times 10$
$= 12 \quad \times 10$
$= 120$

 1

Let's think about how to calculate 40×30.

$$40 \times 30 = 4 \times 10 \times 3 \times 10$$
$$= 4 \times 3 \quad \times 10 \times 10$$
$$= \boxed{} \times \boxed{}$$
$$= \boxed{}$$

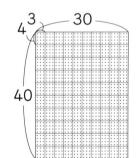

Way to see and think

You can find the same answer even if the order of multiplication is changed.

Summary

Since 40×30 is 100 times of 4×3, the answer is the same as 4×3 with 00 at the end.

 2

Let's calculate the following.

① 2×30 ② 4×60 ③ 70×30 ④ 80×50

❷ How to multiply (2-digit number) × (2-digit number)

Want to solve

1 There are 12 children playing with origami. Each one is making 23 origami shapes. How many sheets of origami paper do they need altogether?

① Let's write a math expression. ☐ × ☐

② About how many sheets of paper is the answer?

If we distribute 20 sheets of paper for 10 people...

Nanami

Want to explain

③ Hiroto thought how to calculate 23 × 12 as shown below. Let's use the following diagram to explain Hiroto's idea.

 Hiroto's idea

My idea is to separate 12 people into 10 people and 2 people.

$$23 \times 12 \begin{cases} 23 \times 2 = \boxed{} \\ 23 \times 10 = \boxed{} \end{cases}$$

Total ☐

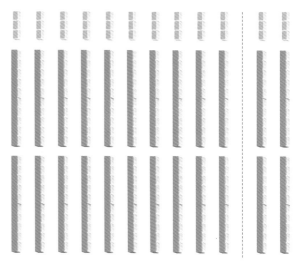

If we decompose 12, we can use the calculation (2-digit number)×(1-digit number).

Can we calculate in vertical form even if we change the multiplier from 1-digit number to 2-digit number?

Daiki

Yui

Way to see and think

You can consider the calculation in the same as you learned before if you decompose the multiplier.

🎯 **Purpose** Can we also calculate (2-digit number)×(2-digit number) in vertical form?

④ Let's think about how to calculate 23×12 in vertical form.

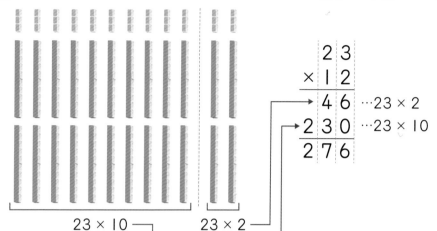

```
   2 3
 × 1 2
 ─────
   4 6  …23 × 2
 2 3 0  …23 × 10
 ─────
 2 7 6
```

23×10 ── 23×2 ──

Way to see and think

It is calculated by decomposing the multiplier into tens and ones.

Multiplication algorithm for 23×12 in vertical form

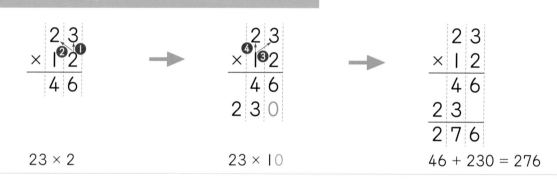

```
   2 3
 × 1 2
 ─────
   4 6
```
23×2

```
   2 3
 × 1 2
 ─────
   4 6
 2 3 0
```
23×10

```
   2 3
 × 1 2
 ─────
   4 6
 2 3
 ─────
 2 7 6
```
$46 + 230 = 276$

🌷**Summary**

We can multiply (2-digit number)×(2-digit number) in vertical form by multiplying each place value, too.

 1 Let's calculate the following in vertical form.

①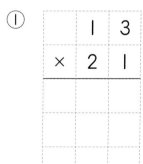

```
     1 3
 ×   2 1
 ───────
```

②

```
     4 1
 ×   1 2
 ───────
```

③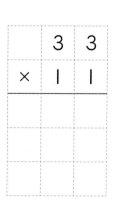

```
     3 3
 ×   1 1
 ───────
```

2 Let's think about how to calculate the following in vertical form.

① 26 × 23

```
      2  6
   ×  2  3
      7  8   ← 26 × 3
   5  2      ← 26 × 20
             ← 26 × 23
```

② 18 × 27

```
      1  8
   ×  2  7
             ← 18 × 7
             ← 18 × 20
             ← 18 × 27
```

What does 52 mean?

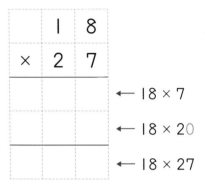

Want to confirm

2 ▶ Let's calculate the following in vertical form.

① 16 × 24 ② 27 × 32 ③ 36 × 23 ④ 17 × 57

Want to think

3 ▶ Let's think about how to calculate the following in vertical form.

① 58 × 46

```
      5  8
   ×  4  6
             ← 58 × 6
             ← 58 × 40
             ← 58 × 46
```

② 37 × 63

```
      3  7
   ×  6  3
             ← 37 × ☐
             ← 37 × ☐
             ← 37 × 63
```

Want to confirm

4 ▶ Let's calculate the following in vertical form.

① 38 × 57 ② 23 × 68 ③ 29 × 44 ④ 28 × 49

Activity

3

Nanami and Daiki thought about how to calculate 35×70 in vertical form as shown below. Let's explain how the children multiplied in vertical form.

Nanami's idea

```
  35          35          35
× 70    →   × 70    →   × 70
  00          00          00
            245         245
                        ┌───┐
                        │   │
                        └───┘
```

Daiki's idea

```
  35          35
× 70    →   × 70
245         2450
```

5 Yui calculated 80×42 as shown below. Let's explain how she improved the calculation.

Yui's idea
```
   42
 × 80
 3360
```

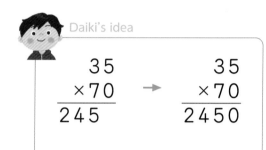

6 Let's calculate the following in vertical form.

① 38×40 ② 75×80 ③ 60×65 ④ 70×25

7 We are going to buy 20 memo pads with a price of 98 yen each. How many yen is the total cost?

Want to solve

1

In class, each student contributes 123 yen to buy materials for arts and crafts. When there are 32 students in class, how many yen will be collected in total?

① Let's write a math expression. ☐ × ☐

② About how many yen is the answer?

③ Let's think about how to calculate.

$$123 \times 32 \begin{cases} 123 \times 2 = \boxed{} \\ 123 \times \boxed{} = \boxed{} \end{cases}$$

Total ☐

Way to see and think

Consider to decompose the multiplier in the same way as in the calculation of (2-digit number) × (2-digit number).

④ Let's explain how to calculate in vertical form.

```
  1 2 3            1 2 3            1 2 3
×   3 2     →    ×   3 2     →    ×   3 2
  2 4 6            2 4 6            2 4 6
                 3 6 9 0          3 6 9
                                  3 9 3 6
```

123 × ☐   123 × ☐   246 + 3690 = ☐

Want to confirm

 Let's calculate the following in vertical form.

① 423 × 21   ② 222 × 43   ③ 315 × 31   ④ 243 × 24

2 We bought a set of snacks with a price of 415 yen for each student. There are 32 students in the class. How many yen is the total cost?

Sale 415 yen

① Let's write a math expression.

② About how many yen is the answer?

③ Let's calculate in vertical form.

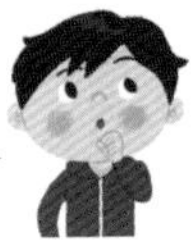

The cost for 10 students is 4150 yen, so...

Daiki

2 Let's calculate the following in vertical form.

① 279 × 64   ② 587 × 57   ③ 754 × 45

④ 680 × 48   ⑤ 368 × 32   ⑥ 294 × 41

3 Yui calculated 508 × 40 as shown on the right.

If there are mistakes, let's explain the reasons. Also, let's calculate correctly.

Yui's idea

```
    5 0 8
 ×   4 0
  2 3 2 0
```

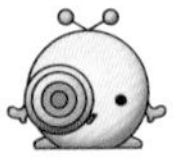

Way to see and think

We can notice a mistake by calculating 500 × 40.

Pay attention to the number of 0s when you do multiplications that involve the number 0.

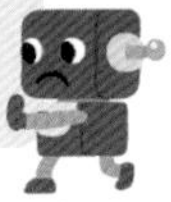

4 Let's calculate the following in vertical form.

① 608 × 50   ② 503 × 60   ③ 205 × 74   ④ 400 × 37

4 Mental calculation

Want to try

1

One bag contains 25 candies. There are 4 bags. How many candies are there altogether?

① Let's write a math expression. [   ]

② Let's think about how to calculate mentally.

> $20 \times 4 = 80,$
> $5 \times 4 = 20,$ so
> $80 + 20 = 100.$

Daiki

Want to confirm

 1 Let's think about how to calculate 250×4 mentally.

> If we think about how many sets of 10 there are in 250...

Hiroto

> If we think based on the calculation of 25×4...

Yui

It is useful to memorize and utilize multiplications that become simple numbers such as $25 \times 4 = 100$.

Want to explain

Activity

2 Hiroto calculated $40 \times 7 \times 25$ mentally as shown below. Let's explain Hiroto's idea.

Hiroto's idea

$$40 \times 7 \times 25 = 40 \times 25 \times 7$$
$$= 1000 \times 7$$
$$= 7000$$

Way to see and think

Using the rule that says that even when changing the order of multiplication, the answer is the same.

Want to confirm

3 Let's calculate the following mentally.

① $2 \times 12 \times 5$ ② $50 \times 37 \times 2$

What you can do now

☐ Understanding how to multiply in vertical form.

1 Let's summarize how to calculate 45×63 in vertical form.

```
      4 5
   ×  6 3
   ─────────
    1 3 5  ← Ⓐ
    2 7 0  ← Ⓑ
   ─────────
    2 8 3 5
```

① Add the answers of 45×3 and $45 \times \boxed{}$.

② Ⓐ is from the multiplication of $\boxed{} \times \boxed{}$.

③ Ⓑ is from the multiplication of $\boxed{} \times \boxed{}$,

and it represents 270 sets of $\boxed{}$.

☐ Can multiply by 2-digit numbers in vertical form.

2 Let's calculate the following in vertical form.

① 5×20 ② 60×30 ③ 40×50

④ 22×14 ⑤ 19×31 ⑥ 27×28

⑦ 43×68 ⑧ 34×86 ⑨ 67×58

⑩ 73×47 ⑪ 25×84 ⑫ 30×92

⑬ 314×21 ⑭ 438×16 ⑮ 593×68

☐ Can make a multiplication expression and find the answer.

3 Let's answer the following problems.

① You need 43 sheets of paper to make each essay collection. When you make 38 essay collections, how many sheets of paper do you need altogether?

② You will buy 18 sets of snacks with a price of 148 yen each and one bottle of tea with a price of 400 yen. If you pay with a 5000-yen bill, how many yen is the change?

Supplementary Problems ⟩ p.149

Usefulness and efficiency of learning

1 Let's find the mistakes in the following calculations in vertical form and calculate correctly.

Understanding how to multiply in vertical form.

①
```
      9 6
  ×   2 8
  -------
    7 6 8
    1 8 2
  -------
    9 5 0
```

②
```
    4 0 8
  ×   6 5
  -------
    2 4 0
    2 8 8
  -------
    3 1 2 0
```

③
```
    2 7 0
  ×   4 6
  -------
    1 6 2 0
    1 0 8
  -------
    2 7 0 0
```

 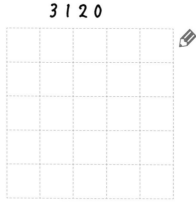 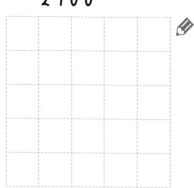

2 Let's make calculations in vertical form by filling in each ☐ with the appropriate numbers.

Can multiply by 2-digit numbers in vertical form.

①

②

Can make a multiplication expression and find the answer.

3 Let's answer the following problems.

① There are 34 students in Yamato's class. He bought a pen with a price of 108 yen for each student in the class. How many yen was the total cost?

② There are 8 bags of cookies in a box. Each bag contains 5 cookies. When you buy 4 boxes, how many cookies are there altogether?

Let's deepen.

43 × 68 and 34 × 86 have the same answer. Is there any rule?

Daiki

Deepen.

Multiplication magic!

We learned how to multiply in vertical form in 3rd grade.

Let's think about the structure of calculations in vertical form.

Want to think

Hiroto explained that 43×68 and 34×86 have the same answer by using calculations in vertical form as shown below.

Hiroto's idea

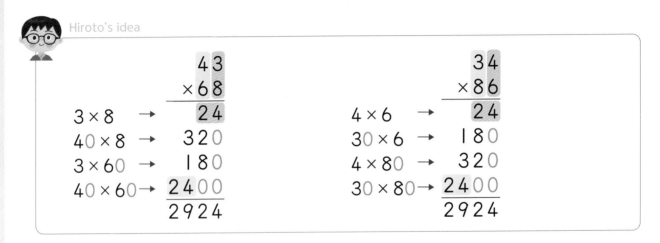

① Hiroto says that both multiplications have the same answer because there are calculations with the same answer in the processes such as $4 \times 6 = 24$ and $3 \times 8 = 24$. Let's think about Hiroto's explanation.

② Let's search for multiplications with 2-digit numbers that have the same answer.

Which pairs of numbers should have the same answer?

What is the length of the remaining part?

1
How long is the width of the blackboard? Let's try to use the length of the tape.

2
As we measure it with a 1-m ruler, there is a remaining part.

How should we represent the length of the remaining part in m?

Remaining part

3
When we divide 1m into 10 equal parts and measure, we can represent it as a decimal number by sets of 0.1.

Remaining part

But we cannot express this part as a number out of 10 equal parts.

4
Is there any other representation method when we cannot represent it as a decimal number?

Can it be represented as a fraction?

Problem Can we represent the length of the remaining part as some form by not using a decimal number?

16 Fractions

Let's explore how to represent the size of a number by measuring and dividing and its structure.

1 Fractions

Want to know Representing length as a fraction

1 We measured the width of the blackboard by cutting the tape and measuring it with a 1-m ruler. The length is 1 m and a remaining part. Let's think about how many m the length of the remaining part is.

1-m ruler

Remaining part

 Nanami: When we divide into 10 equal parts and measure, we can represent it as a decimal number. But...

Hiroto: We represented the size of one part out of 4 equal parts as a fraction.

Purpose Can we represent the length of the remaining part as a fraction?

① Let's compare the length of the remaining part with the length of one part of the 1-m tape that has been divided into 2, 3, and 4 equal parts.

1 m

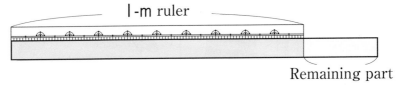

The length of the remaining part is equal to one part of 1 m that has been divided into 4 equal parts.

We learned in 2nd grade that one part of an object that has been divided into 4 equal parts is represented as $\frac{1}{4}$ of the object.

86

The length of one part resulting from dividing **l m** into 4 equal parts is called "**one fourth meter**" or "**one over four meter**," and is written as $\frac{1}{4}$ **m .**

$$\frac{1}{4} \begin{matrix} ❸ \\ ❶ \\ ❷ \end{matrix}$$

Want to think

② How many pieces of the remaining part make **l m?**

l m

l piece of the remaining part

2 pieces of the remaining part

3 pieces of the remaining part

4 pieces of the remaining part

Summary

The length of the remaining part for which 4 pieces are equal to **l m** is the same as one part of **l m** equally divided into 4 parts. The length of the remaining part is $\frac{1}{4}$ m.

Way to see and think

It is represented with fractions by using how many pieces of the remaining part there are.

Want to confirm

1 How many **m** are the following lengths?

① The length of one part that is made by dividing **l m** into 3 equal parts.

l m

☐ m

② The length of the remaining part for which 3 pieces are equal to **l m.**

l m   Remaining part

③ The length of one part that is made by dividing **l m** into 5 equal parts.

l m

☐ m

④ The length of the remaining part for which 2 pieces are equal to **l m.**

l m   Remaining part

Let's color each of the parts that apply to the given lengths.

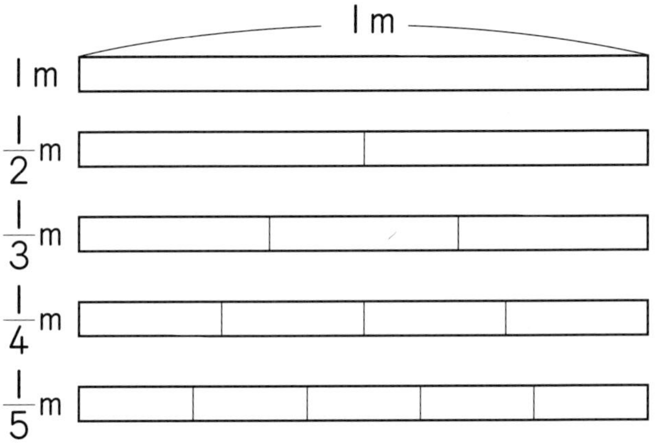

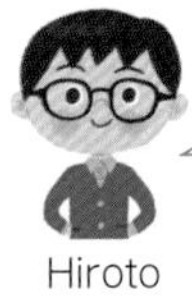

1 m

$\frac{1}{2}$ m

$\frac{1}{3}$ m

$\frac{1}{4}$ m

$\frac{1}{5}$ m

Hiroto: They look like rulers to measure length with fractions. I want to measure various things with these rulers.

2 ▷ Let's discuss how many m the length of the tape shown on the right is.

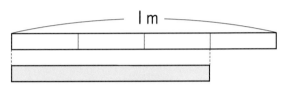
1 m

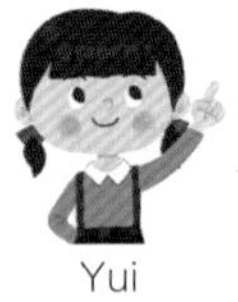

Yui: It is the same as the length of 3 parts of 1 m that has been divided into 4 equal parts.

Daiki: The length for 1 part is $\frac{1}{4}$ m.

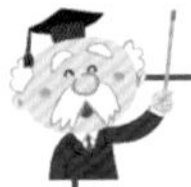

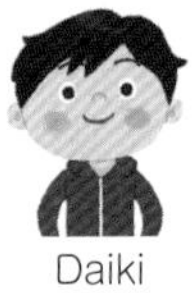

The length of 3 sets of $\frac{1}{4}$ m is written as $\frac{3}{4}$ m and read as "**three fourths meter**" or "**three over four meter**."

Way to see and think

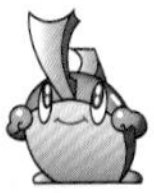

The idea is to consider $\frac{1}{4}$ m as the unit.

3 ▷ How many m are the lengths of the following colored parts?

Way to see and think

You should consider the size of one unit.

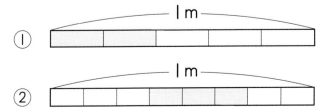

① 1 m

② 1 m

Want to represent

 4 Let's think about how to represent the amount of water shown on the right.

① How many L is the amount of water for one scale?

② How many L is the amount of water?

Numbers such as $\frac{1}{3}$, $\frac{3}{4}$, and $\frac{4}{5}$ are called **fractions**. The number above the bar is called the **numerator** and the number below the bar is called the **denominator**.

$\frac{3\cdots\text{Numerator}}{4\cdots\text{Denominator}}$

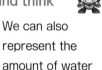

Way to see and think

We can also represent the amount of water with fractions in the same way as with length.

Want to think

 5 When 1 L of milk is equally divided among 3 children, how many L is the amount of milk for 2 children? What do the denominator and numerator represent?

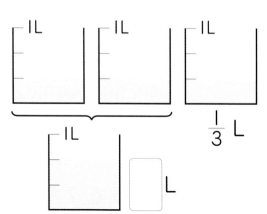

The denominator represents the number of equal parts of the original quantity, such as 1 m or 1 L, and the numerator represents the number of gathered parts.

Want to confirm

 6 How many L are the following amounts of water?

① ② ③

Way to see and think

You should consider how many equal parts are divided.

7 Let's color each of the parts that apply to the given length and amount of water.

① $\frac{5}{6}$ m

— I m —

② $\frac{3}{5}$ L

IL

That's it.

Let's measure by using fractions.

0m	$\frac{1}{4}$m	$\frac{2}{4}$m	$\frac{3}{4}$m	Im

① Let's make rulers to measure length with fractions by dividing a I-m tape into equal parts. Let's make rulers to measure length with fractions with denominators of 3, 5, 6, 7, 9, and 10, and then measure the length of different objects.

How to make a ruler for fractions with a denominator of 9

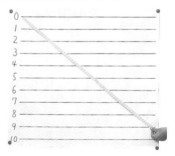

② Let's make a fraction measuring cup by placing a scale for fractions on a I-L measuring cup.

How to make a scale for fractions with a denominator of 7

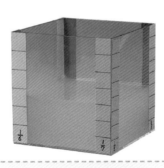

❷ Structure of fractions

Size of fractions

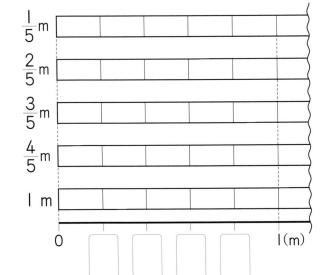

1 Using the diagram shown on the right, let's color each of the parts that apply to the given length from left to right.

① Let's fill in each ☐ with the appropriate numbers.

② How many sets of $\frac{1}{5}$ m are there in $\frac{3}{5}$ m?

③ How many sets of $\frac{1}{5}$ m are there in 1 m?

Want to confirm

 How many L are 6 sets of $\frac{1}{6}$ L?

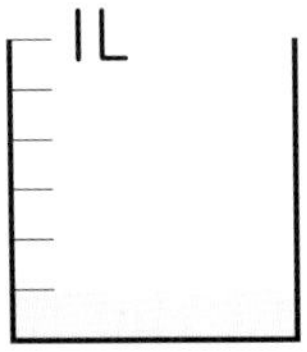

Fractions whose numerators and denominators are the same, such as $\frac{5}{5}$ and $\frac{6}{6}$, are equal to 1.

$$\frac{6}{6} = 1$$

Want to try

 Which is longer, $\frac{3}{5}$ m or $\frac{4}{5}$ m?

 Let's fill in each ☐ with the appropriate inequality sign.

① $\frac{3}{4}$ m ☐ $\frac{2}{4}$ m   ② $\frac{5}{7}$ L ☐ $\frac{6}{7}$ L

③ $\frac{7}{8}$ dL ☐ 1 dL   ④ $\frac{8}{9}$ m ☐ 1 m

Way to see and think

We can think how many sets of $\frac{1}{5}$ make each fraction.

2 How should we represent lengths that are longer than 5 sets of $\frac{1}{5}$ m? Let's discuss.

> It represents a number larger than 1.

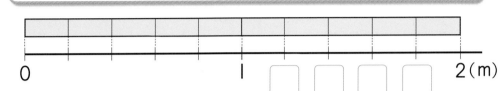

Yui

The length of 6 sets of $\frac{1}{5}$ m is written as $\frac{6}{5}$ m and read as "**six fifths meter**" or "**six over five meters**."

① How many m are 7, 8, and 9 sets of $\frac{1}{5}$ m?

② How many m are 10 sets of $\frac{1}{5}$ m?

> We can represent it as both a fraction and a whole number.

Hiroto

4 Let's represent the following numbers on the number line with an ↓.

① $\frac{1}{6}$m ② $\frac{3}{6}$m ③ $\frac{8}{6}$m ④ $\frac{6}{6}$m

0   1   2(m)

5 Let's write the numbers that are represented by an ↑ on the following number line.

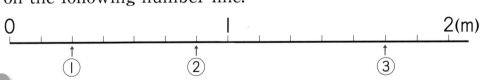

Way to see and think

First, you should think about the size of each scale on the number line. Then, consider how many of this scales each number is.

6 Let's fill in each ☐ with the appropriate inequality sign.

① $\frac{5}{4}$m ☐ $\frac{4}{4}$m ② $\frac{7}{8}$L ☐ $\frac{8}{8}$L ③ $\frac{5}{3}$m ☐ 2m

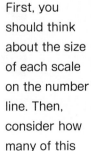

3 Let's fill in each ☐ with the appropriate fraction above the number line and each ☐ with the appropriate decimal number below the number line.

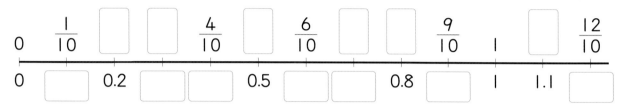

① $\frac{1}{10}$ is 1 set of 1 divided by how many equal parts?

② 0.1 is 1 set of 1 divided by how many equal parts?

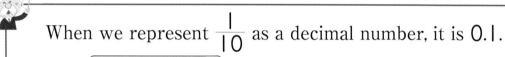

When we represent $\frac{1}{10}$ as a decimal number, it is 0.1.

$$\frac{1}{10}=0.1$$

The tenths place is also called " $\frac{1}{10}$ **place**."

0.8
···Ones place
···Tenths place
··· $\frac{1}{10}$ place

Want to confirm

7 Let's answer the following problems.

① Let's represent $\frac{1}{10}$ as a decimal number.

② Let's represent 0.7 as a fraction.

③ Let's represent 5 sets of $\frac{1}{10}$ as a fraction and as a decimal number.

Want to try

8 Let's fill in each ☐ with the appropriate equality or inequality sign.

① $\frac{3}{10}$ ☐ 0.2 ② 0.1 ☐ $\frac{1}{10}$ ③ 0.7 ☐ $\frac{8}{10}$

Way to see and think

It's easy to compare numbers by aligning fractions into decimal numbers or decimal numbers into fractions.

1 Daiki drank $\frac{1}{5}$ L of milk yesterday and $\frac{2}{5}$ L of milk today. How many L of milk did he drink altogether?

① Let's write a math expression.

Is it an addition bacause of the word "altogether"?

Yui

Can we add in the same way as with decimal numbers?

Hiroto

Purpose   Can we also add fractions?

② Let's think about how to calculate.

$\frac{1}{5} + \frac{2}{5} =$ □

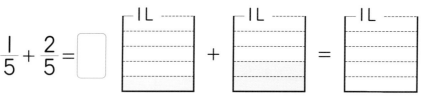

□ sets of $\frac{1}{5}$L   □ sets of $\frac{1}{5}$L   □ sets of $\frac{1}{5}$L

Way to see and think

For the total, you should consider how many sets of $\frac{1}{5}$ there are.

Summary

For the addition of fractions, by considering how many sets of the fraction as one unit there are, add the numerators to find the answer.

1 Let's find the answer of $\frac{3}{5} + \frac{2}{5}$ as both a fraction and a whole number.

2 Let's calculate the following.

① $\frac{2}{7} + \frac{4}{7}$ ② $\frac{1}{8} + \frac{5}{8}$ ③ $\frac{3}{4} + \frac{1}{4}$ ④ $\frac{7}{10} + \frac{3}{10}$

Way to see and think

What kind of number does it become when the numerator and denominator are equal?

2 From a $\frac{7}{8}$ m tape, $\frac{5}{8}$ m was cut off. How many m were left?

① Let's write a math expression.

Daiki

Should we subtract to find the remaining section?

Since we can add, it seems we can also subtract.

Nanami

Purpose Can we subtract fractions in the same way as adding fractions?

② Let's think about how to calculate.

$$\frac{7}{8} - \frac{5}{8} = \boxed{}$$

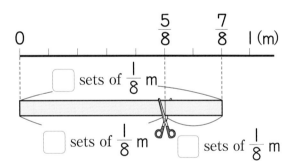

$\boxed{}$ sets of $\frac{1}{8}$ m

$\boxed{}$ sets of $\frac{1}{8}$ m $\boxed{}$ sets of $\frac{1}{8}$ m

Way to see and think

For the remaining section, you should consider how many sets of $\frac{1}{8}$ there are.

Summary

For the subtraction of fractions, by considering how many sets of the fraction as one unit there are, subtract the numerators to find the answer.

Want to confirm

3 Let's find the answer of $1 - \frac{5}{8}$.

4 Let's calculate the following.

Way to see and think

You should think how many sets of $\frac{1}{8}$ there are in 1.

① $\frac{3}{4} - \frac{1}{4}$ ② $\frac{5}{6} - \frac{3}{6}$ ③ $1 - \frac{3}{7}$ ④ $1 - \frac{1}{10}$

What you can do now

☐ Understanding the structure of fractions.

1 Let's fill in each ☐ with the appropriate numbers.

① $\frac{3}{5}$ dL is ☐ sets of $\frac{1}{5}$ dL .

② $\frac{☐}{☐}$ m is 5 sets of $\frac{1}{6}$ m .

③ ☐ sets of $\frac{1}{8}$ L are $\frac{3}{8}$ L .

④ 5 sets of $\frac{1}{5}$ cm are ☐ cm.

☐ Understanding the meaning of fractions.

2 Let's color each of the parts that apply to the given lengths and amounts of water.

① $\frac{3}{4}$ m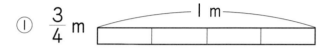

② $\frac{5}{7}$ m

③ $\frac{2}{3}$ L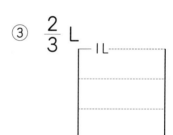

④ $\frac{3}{5}$ L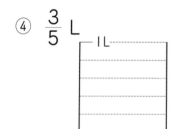

⑤ $\frac{4}{6}$ L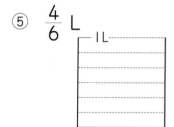

☐ Understanding the relationship between fractions, decimal numbers, and whole numbers.

3 Let's fill in each ☐ with the appropriate equality or inequality sign.

① $\frac{2}{3}$ ☐ $\frac{1}{3}$

② $\frac{5}{8}$ ☐ $\frac{7}{8}$

③ 1 ☐ $\frac{3}{4}$

④ $\frac{6}{10}$ ☐ 0.6

⑤ $\frac{9}{9}$ ☐ 1

⑥ $\frac{5}{10}$ ☐ 0.4

☐ Can add and subtract fractions.

4 Let's calculate the following.

① $\frac{1}{4} + \frac{3}{4}$

② $\frac{2}{8} + \frac{4}{8}$

③ $\frac{3}{5} + \frac{1}{5}$

④ $\frac{6}{7} - \frac{3}{7}$

⑤ $\frac{5}{6} - \frac{4}{6}$

⑥ $1 - \frac{1}{3}$

Supplementary Problems ▸ p.150

Usefulness and efficiency of learning

1 Let's fill in each ☐ with the appropriate fraction above the number line and each ☐ with the appropriate decimal number below the number line.

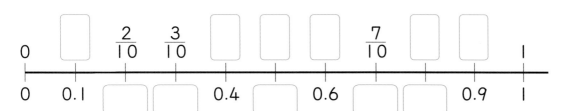

☐ Understanding the structure of fractions.

☐ Understanding the meaning of fractions.

☐ Understanding the relationship between fractions, decimal numbers, and whole numbers.

2 Using the cards numbered from 1 to 5 as numerators, let's make fractions with 5 as the denominator.

☐ Understanding the structure of fractions.

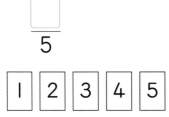

① Let's make a fraction such that 3 sets of it is equal to $\frac{3}{5}$.

② Let's make a fraction that is equal to 1.

③ Let's make fractions that are smaller than $\frac{4}{5}$.

④ Let's make a fraction that is larger than $\frac{3}{5}$ but smaller than 1.

☐ Can add and subtract fractions.

3 There is a 1 m ribbon. We gave $\frac{1}{8}$ m of it to Arisa and $\frac{5}{8}$ m of it to Yurina. As for the ribbon, how many m are left?

Let's deepen.

What is the difference between $\frac{1}{4}$ learned in 2nd grade and $\frac{1}{4}$ m?

Hiroto

Utilize in mathematics.

Deepen.

Is it the same length?

Want to find

Nanami and Daiki are discussing $\frac{1}{4}$ m. Which idea is correct?

Nanami

Since $\frac{1}{4}$ m is the length of one part that is made by dividing 1 m into 4 equal parts, the colored part is $\frac{1}{4}$ m.

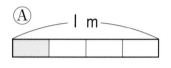

Ⓐ — 1 m —

Daiki

I think this is $\frac{1}{4}$ m because one tape is divided into 4 equal parts.

Ⓑ — 2 m —

Way to see and think

You should think about what is one unit.

① How many equal parts are there in 1 m for tape Ⓐ and tape Ⓑ respectively?

② Let's discuss why the colored part of tape Ⓑ is longer than the colored part of tape Ⓐ.

Is it okay to represent different lengths in the same way?

Hiroto

$\frac{1}{4}$ m is the length of one part that is made by dividing 1 m into 4 equal parts.

In tape Ⓑ, 2 m are divided into 4 equal parts. So, we cannot say the length of one part is $\frac{1}{4}$ m.

Each "one part" is different based on the length of the original tape.

Yui

Nanami

Daiki

Want to deepen

Let's think about how many m is the colored part of tape Ⓑ.

Weight found in town

 Problem

How should we explore weight?

17 Weight
Let's explore how to represent weight and its structure.

Want to know How to compare weight

1 Which is the heaviest, the scissors, the glue stick, or the compass? Also, which is the lightest?

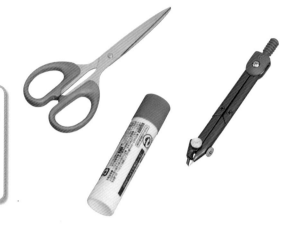

 Daiki: I wonder if we can compare weight by holding objects in our hands.

Yui: Can we compare by hanging each object with a rubber?

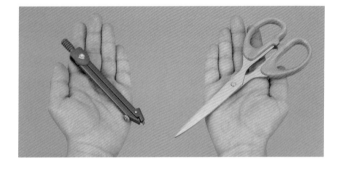

① A balance is a device for comparing the weight of objects. The following pictures show the result after comparing the weight of the objects. Let's arrange from the heaviest to the lightest.

Way to see and think

Can you explain the reason of this order?

2 How should we represent the weight of the scissors, the glue stick, and the compass by using numbers?

Nanami

We can explore how many erasers or pencils correspond to the weight of each object.

We represented length and amount of water using numbers.

Hiroto

Purpose Can we represent the weight of objects by using numbers?

① Let's represent the weight by using one-yen coins.

Scissors　44 pieces of one-yen coins

Object	Number of one-yen coins
Scissors	
Compass	
Glue stick	

Way to see and think

The weight of each object is represented using a number by how many one-yen coins there are.

Summary

We can measure weight by how many sets of the unit of weight there are.

There is a unit called **gram** that is used to measure weight. I gram is written as I g. The weight of a one-yen coin is I g.

② How many g is the weight of the scissors, the glue stick, and the compass respectively?

Let's measure the weight of objects in our surroundings by using one-yen coins.

3 We use scales to measure weight. Let's measure the weight of a pencil case by using the scale.

g is the same as g.

0 50 100 150 200 250 300

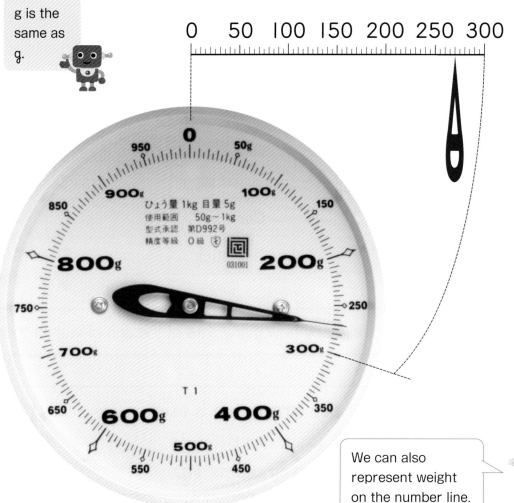

We can also represent weight on the number line.

Yui

Way to see and think

How many parts are the sections between 0 and 50 divided into?

① Up to how many g can the scale weigh?

② How many g does the smallest scale represent?

③ How many g is the weight of the pencil case?

102

2 ▶ Let's read the weight shown on the following scales.

①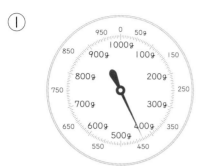

②

3 ▶ The weight of a paint box is **870 g.**
Let's draw the needle on the scale shown
on the right representing this weight.

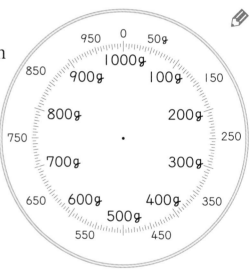

 Let's avoid placing very heavy things on the tray.

How to use a scale

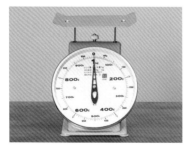

(1) Put the scale on a flat surface.

(2) At first, make sure the needle is at 0.

(3) Stay in front of the scale to read properly.

4 Let's measure the weight of a tape cutter by using a scale.

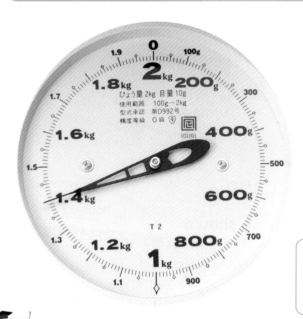

How should we represent weight when the object weighs more than 1000 g?

Yui

1000 g is called 1 **kilogram** and is written as 1kg.

| 1 kg = 1000 g |

1kg

Way to see and think

1 kg is the same weight as 1000 times of 1g.

① Up to how many kg can the scale weigh?

② How many g does the smallest scale represent?

③ How many kg and g is the weight of the tape cutter?

1 kg 400 g is also read as "1 kilo and 400 grams."

1 L of water weighs 1 kg.

This is similar to the relationship between km and m when measuring length.
1 km = 1000 m

Daiki

4 How many one-yen coins are needed to make 1kg?

5 Let's make various things that have a weight of 1kg.

6 How many kg and g is the weight shown on the following scales? Also, how many g?

① 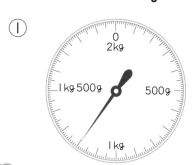   ②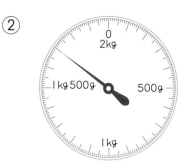

7 The calligraphy bag weighs 1kg 600g. Let's draw the needle on the scale shown on the right representing this weight.

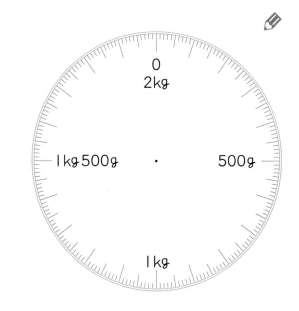

8 Let's fill in each ☐ with the appropriate numbers.

① 3kg = ☐ g   ② 2kg 500 g = ☐ g

③ 1800 g = ☐ kg ☐ g

Want to explore

5 Which scale should we use to measure the weight of the following objects? Let's choose a scale from ⓐ, ⓑ, and ⓒ.

Way to see and think

We need to estimate the weight of each object.

① Watermelon   ② Textbook   ③ Your weight

ⓐ    ⓑ    ⓒ

Want to confirm

9 Let's measure the weight of various objects by using a scale.

Let's estimate at first.

Exploring the weight

Object	Estimated weight	Actual weight
Kanji dictionary		

Want to know

10 The symbol "3t" on the sign post represents weight. In what kind of place is "t" used?

Besides g and kg, there is another unit of weight called **ton** (metric ton). The weight 1000 kg is called 1 tonne and is written as 1t.

1t = 1000 kg

Want to explore

1 Let's explore the relationship between the units of length, amount of water, and weight.

Length	mm, cm, m, km
Amount of water	mL, dL, L
Weight	g, kg, t

Way to see and think

Are there any rules for the units with prefixes k (kilo) and m(milli)?

① Let's fill in each ☐ with the appropriate numbers. Also, let's discuss the noticed things.

1 kg = ☐ g   1 km = ☐ m

1 m = ☐ mm   1 L = ☐ mL

Hiroto: Can the prefix "milli" be used for the unit of weight, in the same way as for the units of length and amount?

Nanami: It looks like there is kL for the amount of water.

$$1000mg = 1 g \qquad 1000L = 1kL$$

	milli m	centi c	deci d			kilo k			
Length	1mm	1cm		1m		1km			
Amount of water	1mL		1dL	1L		1kL			
Weight	1mg			1g		1kg			1t

10 times   100 times   1000 times   1000 times

1000 times

Want to try

1 Yui utilized the structure of the units of weight to make a unit conversion table as shown on the right. Let's make unit conversion tables for length and amount of water in the same way.

t			kg			g
			3	0	0	0

If we write 3 under kg and place zeros up to the place of g, we can understand that 3kg = 3000g.

3 Weight represented with decimal numbers

Want to deepen

1 After measuring Keita's weight, his weight is 31.8 kg. How many kg and g does this represent?

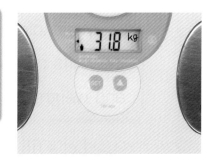

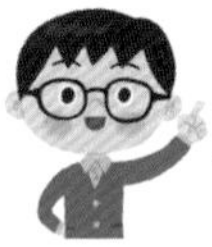

Hiroto

0.1 is one set after 1 has been divided into 10 equal parts.

1 kg is 1000 g. If we divide it into 10 equal parts, then...

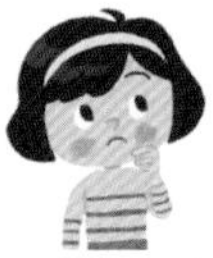

Nanami

Purpose How should we read weight when it is represented with decimal numbers?

① How many g is 0.1 kg?

② How many kg and g is 31.8 kg?

Way to see and think

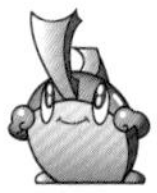

We should think about one set after 1000 g is divided into 10 equal parts.

Summary

When using decimal numbers, weight can be represented with one unit of quantity such as kg.

Want to confirm

1 How many kg and g are the following weights?

① 4.8 kg　② 21.2 kg　③ 15.7 kg

Want to try

2 How many kg are the following weights?

① 8 kg 300 g　② 14 kg 800 g　③ 10 kg 100 g

1 The weight of a lump of clay was measured after the shape was changed several times. What happened to its weight?

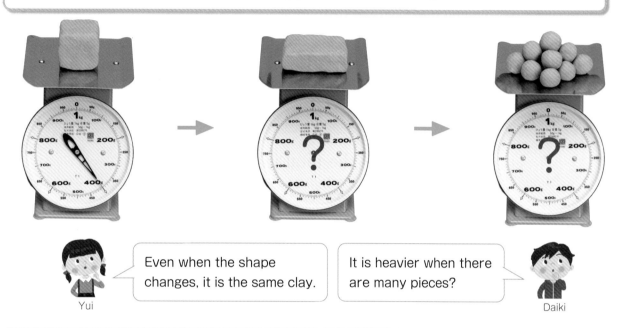

Even when the shape changes, it is the same clay.

Yui

It is heavier when there are many pieces?

Daiki

The weight of the amount of an object is the same even when the shape changes.

1 There are iron, aluminum, rubber, and wood blocks with the same size. Is the weight the same? Let's estimate and compare.

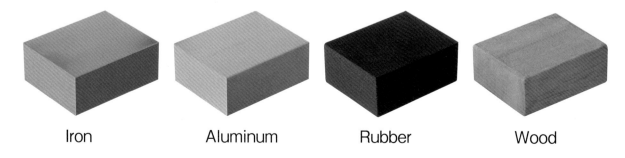

Iron　　　　　Aluminum　　　　Rubber　　　　Wood

Different materials may differ in weight even if they have the same size.

❺ Calculation of weight

1 There are 900 g of oranges placed in a basket that weighs 400 g. How many g is the total weight?

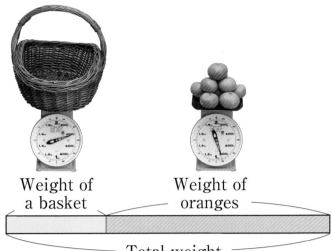

Weight of a basket   Weight of oranges

Total weight

① How many g is the total weight?

   400 g + 900 g

② How many kg and g is this?

Way to see and think

Can we calculate weight in the same way as length and amount of water?

 A school bag weighs 900 g. The total weight of the school bag with books and notebooks is 3 kg 200 g. How many kg and g is the weight of the books and notebooks?

What kind of calculation should we do to find the weight of the books and notebooks?

Hiroto

2 Let's find the following weights.

① Erina's weight is 24 kg and Reo's weight is 26 kg. How many kg is the total weight of Erina and Reo?

② Arata weighed 3200 g at birth and weighed 9100 g on his first birthday. How many g did his weight increase in 1 year?

What you can do now

☐ Understanding how to represent weight.

1 Let's summarize about how to represent weight.

① The units that represent weight are [   ] , [   ] , and t.

② The relationship between these units is 1kg=1000 [   ] .

③ The weight of a one-yen coin is [   ] .

④ The weight of 1 L of water is [   ] .

☐ Can read weight by using various scales.

2 How many kg and g is the weight in ① and ②? Also, how many g?

①

②

[   ] kg [   ] g
[   ] g

[   ] kg [   ] g
[   ] g

☐ Understanding the units of weight.

3 Let's fill in each [   ] with the appropriate numbers.

① 4300 g= [   ] kg [   ] g   ② 6 kg= [   ] g

③ 800 g= [   ] kg   ④ 2.7 kg= [   ] g

☐ Can calculate weight.

4 A 750 g book is placed inside a 600 g box. How many kg and g is the total weight? Also, how many g is the difference in weight between the book and the box?

Supplementary Problems p.152

Usefulness and efficiency of learning

1 Let's fill in each ☐ with the appropriate numbers.

① 1 kg 350 g = ☐ g

② 3074 g = ☐ kg ☐ g

③ 0.6 t = ☐ kg

④ 10.1 kg = ☐ g

⑤

	milli m	centi c	deci d			kilo k		
Length	1mm	1cm		1m		1km		
Amount of water	1mL		1dL	1L		1kL		
Weight	1mg			1g		1kg		1t

10 times 100 times

☐ times

☐ times   ☐ times

☐ Understanding how to represent weight.

☐ Understanding the units of weight.

2 How many kg and g is the weight shown in ①, ②, and ③? Also, how many g?

①

②

③

☐ Can read weight by using various scales.

3 Let's answer the following problems.

① After pouring 2 L of water into a goldfish tank, the total weight became 3 kg 200 g. How many kg and g is the weight of the goldfish tank?

② There are 6 cans weighing 350 g each in a basket that weighs 400 g. How many g is the total weight?

☐ Can calculate weight.

Let's deepen.

Can we utilize weight in our daily life?

Daiki

112

Deepen.

Nutrients contained in rice

Rice contains a lot of nutrients that our body requires. The table shown below summarizes the nutrients contained in 100 g of steamed rice. Let's answer the following problems.

Brown rice

Rice after removing the chaff from the rough rice.
It is made of aleurone cell layer, embryo bud, and starch.

Milled rice

Rice after removing the aleurone cell layer and the embryo bud from the brown rice.

Milled rice with embryo

Milled rice left with the embryo bud after milling.

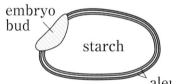

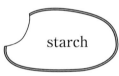

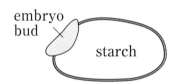

Nutrients contained in rice   (g)

	Moisture	Protein	Fat	Carbohydrate	Others
Steamed brown rice	63	3.3	1.3	31.8	
Steamed milled rice	65	2.6	0.5	31.8	
Steamed milled rice with embryo	65	2.6	0.8	31	

① "Others" consists of other nutrients such as vitamins. How many g is the weight in "Others" for each type of rice?

② Kanon ate 100 g of steamed rice for each meal in one day choosing from the above rice types. In total, she ate 9.2 g of protein and 1.3 g of others after 3 meals. Let's think about the combination of rice she ate in one day.

Problem
What units of quantity do we know?

◎ Let's arrange and read the units of length from the smallest to the largest.

mm → cm → m → km

Let's confirm the relationship between units.

I km = [   ] m

I m = [   ] cm

I cm = [   ] mm

I m = [   ] mm

The units of length have this same symbol.

◎ Let's arrange and read the units of amount of water from the smallest to the largest.

mL → dL → L

Let's confirm the relationship between units.

I L = [   ] dL

I dL = [   ] mL

I L = [   ] mL

The units of amount of water have this same symbol.

Are the units of length based on "m" and the units of amount of water based on "L"?

Hiroto

Are the units of weight based on "g"? But "t" does not have "g."

Daiki

What is the meaning of the symbols found in the units?

Nanami

◎ Let's arrange and read the units of weight from the smallest to the largest.

| g | → | kg | → | t |

Let's confirm the relationship between units.

1 kg = ☐ g

1 t = ☐ kg

The units of weight are not aligned.

◎ Let's gather the units with the same prefix.

length   weight

| km | | kg |

Is there kL as a unit of amount of water?

In both cases, "k" is read as kilo.
What is the meaning of "k"?

length   weight

| mm | | mL |

Is there "mg" as a unit of weight?

In both cases, "m" is read as milli.
What is the meaning of "m"?

I have found "cL" that uses the same prefix as "cm." Here, "c" is read as centi. What is the meaning of "c"?

◎ Let's find units from our surroundings.

Protein	3.9g
F a t	22.7g
Carbohydrate	48.0g
S o d i u m	86mg

2008

MOUSSEAU, PROPRIÉTAIRE À TEUILLAC
TÉ PAR NOS ŒNOLOGUES
R GINESTET À 33360 FRANCE
75 d.

It seems that gathering 1000 sets becomes a bigger unit. I have seen there are so many types of units in our surroundings.

Yui

Want to connect

Are there any other units used in our surroundings?

Yui

How many people were riding the bus?

The number of people inside the bus increased by 4 people.

(Number of people riding the bus) + 4 is the current number of people inside the bus.

There are 17 people inside the bus.

Then, it is (number of people riding the bus) + 4 = 17. Can we write a math sentence without using words?

Problem How should we represent math sentences when there are unknown numbers?

18 Math Sentences Using the □

Let's represent the relationship between numbers as a math sentence using the □.

Want to represent Addition using the □

1

Some strawberries were placed on a basket that weighs 200 g. Altogether, the weight is 600 g. Let's answer the following problems.

① The relationship between the weight of the strawberries, the weight of the basket, and the total weight is represented in the following diagram. Let's fill in the ( ) with the appropriate words.

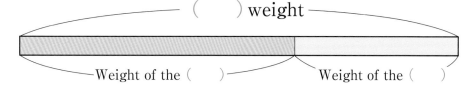

( ) weight

Weight of the ( )   Weight of the ( )

Way to see and think

The relationship between numbers is easier to understand when it is shown in a diagram.

② Let's represent the diagram shown in ① as a math sentence using words for finding the total weight.

$$\boxed{} + \boxed{} = \boxed{}$$

③ Let's represent the above math sentence using words as a math sentence using the □ as an unknown number.

$$\boxed{}$$

When there are unknown numbers, these numbers can be represented by □. Also, we can consider the math expression □+ 200 as a representation of the total weight.

④ Let's think about how to find the number that applies to the ☐.

Daiki

It looks like we can find it by placing numbers into the ☐ in order.

Can we find it using a diagram?

Nanami

 Purpose How can we find the number that applies to the ☐?

Want to compare

⑤ Let's compare the ideas of the children.

 Daiki's idea

I will find the number by placing 100, 200, ... into the ☐ in ☐ + 200 = 600.

$$100 + 200 = 300$$
$$200 + 200 = 400$$
$$300 + 200 = 500$$
$$400 + 200 = 600$$

Nanami's idea

I considered using a diagram.

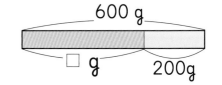

600 g
☐ g   200 g

$$☐ + 200 = 600$$
$$☐ = 600 - 200$$

Summary

We can find the number that applies to the ☐ by changing the addition sentence using the ☐ into a subtraction.

Want to confirm

 1 400 g of apples were placed in a container. Altogether, the weight is 850 g. Let's answer the following problems.

① Let's represent this as an addition sentence using ☐ g as the weight of the container.

② Let's find the number that applies to the ☐.

Want to try

 2 Let's find the number that applies to each ☐.

① $18 + ☐ = 50$ ② $☐ + 27 = 105$

2

At the book store, a book was bought for 380 yen and the change was 120 yen. How much money was paid at the cashier?

① The relationship between the money that was paid, the price of the book, and the change is represented in the following diagram. Let's fill in the (   ) with the appropriate words.

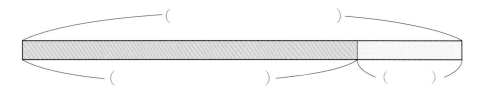

② Let's represent the diagram shown in ① as a math sentence using words for finding the change.

$$\boxed{} - \boxed{} = \boxed{}$$

③ Let's represent this as a subtraction sentence using □ yen as the money that was paid.

$$\boxed{}$$

④ Let's find the number that applies to the □ .

Can we place numbers into the □ to find the answer?
Hiroto

Can we consider using a diagram in the same way as Nanami thought in page 118?

Yui

 3 Let's find the number that applies to each □ .

① □ − 51 = 34   ② □ − 59 = 141

3 10 pencils with the same price were bought for 800 yen. How many yen was the cost for one pencil?

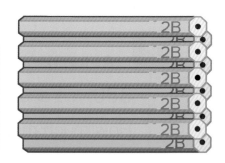

① The relationship between the total cost, the cost for one pencil, and the number of pencils is represented in the following diagram. Let's fill in the ( ) with the appropriate words.

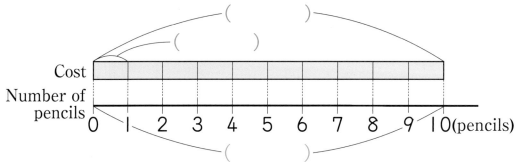

② Let's represent the diagram shown in ① as a math sentence using words for finding the total cost.

[   ] × [   ] = [   ]

③ Let's represent this as a multiplication sentence using □ yen as the cost for one pencil.

④ Hiroto discovered how to find the number that applies to the □ as shown on the right. Let's explain Hiroto's idea.

Hiroto's idea

□ × 10 = 800
□ = 800 ÷ 10
□ = 80

4 When buying 10 m of a ribbon, the total cost was 750 yen.

① Let's represent this as a math sentence for finding the total cost using □ yen as the cost for 1 m of the ribbon.

② Let's find the number that applies to the □.

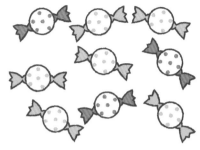

4 The candies, bought at 8 yen for each piece, had a total cost of 72 yen. How many pieces of candies were bought?

① Let's represent this in the following diagram using the □ as an unknown number.

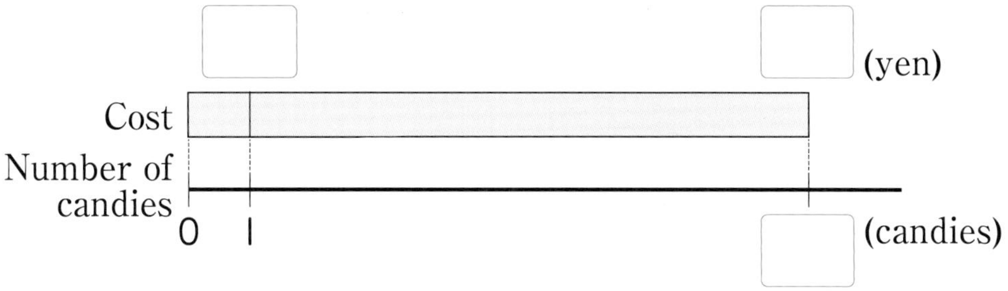

Cost

Number of candies

0   1

(yen)

(candies)

② Let's represent the diagram shown in ① as a math sentence using words for finding the total cost.

③ Let's represent this as a multiplication sentence using □ pencils as the number of candies that were bought.

④ Let's think about how to find the number that applies to the □. Also, let's find the answer.

5 Let's find the number that applies to each □.

① $7 \times \square = 56$ ② $\square \times 4 = 28$

③ $\square \times 6 = 24$ ④ $9 \times \square = 63$

⑤ $10 \times \square = 300$ ⑥ $\square \times 10 = 1000$

6 Let's look at the picture shown on the right and think about a story that becomes into $\square \times 6 = 48$. Also, let's find the number that applies to the □.

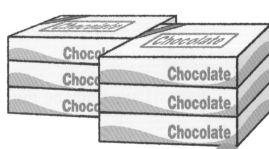

5 Some sheets of colored paper will be equally divided into 6 people, so that each person receives 8 sheets of paper.

① The relationship between the total number of sheets, the number of people, and the number of sheets for each person is represented in the following diagram. Let's fill in the (　) with the appropriate words.

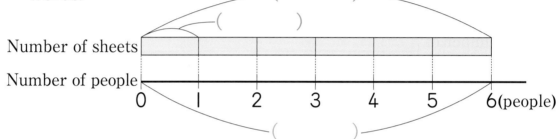

② Let's represent the diagram shown in ① as a math sentence using words for finding the number of sheets for each person.

$$\boxed{} \div \boxed{} = \boxed{}$$

③ Let's represent this as a division sentence using □ sheets as the total number of sheets.

$$\boxed{}$$

④ Let's find the number that applies to the □.

Daiki: The total number of sheets is an unknown number.

We sometimes use multiplication to find the total number.

Nanami

7 Let's find the number that applies to each □.

① $□ \div 4 = 5$　　② $□ \div 7 = 9$

What you can do now

☐ Can represent a situation as a math sentence for finding an unknown number.

1 Let's answer the following problems.

① All the children in the class folded paper cranes. They folded 240 cranes yesterday. Today, they folded more cranes. They made 500 cranes in total. Let's represent this as an addition sentence using ☐ cranes as the number of cranes folded today. Also, let's find the number that applies to the ☐.

Math sentence: ☐   Answer: ☐ cranes

② After buying a chocolate with a cost of 280 yen, the remaining amount of money was 350 yen. Let's represent this as a subtraction sentence using ☐ yen as the original amount of money. Also, let's find the number that applies to the ☐.

Math sentence: ☐   Answer: ☐ yen

③ When buying 10 m of rope, the total cost was 980 yen. Let's represent this as a multiplication sentence using ☐ yen as the cost for 1 m of rope. Also, let's find the number that applies to the ☐.

Math sentence: ☐   Answer: ☐ yen

④ After some jellies were divided among 7 people, each person received 3 jellies. Let's represent this as a division sentence using ☐ jellies as the total number of jellies. Also, let's find the number that applies to the ☐.

Math sentence: ☐   Answer: ☐ jellies

☐ Can find the number that applies to the ☐.

2 Let's find the number that applies to each ☐.

① $\square + 3 = 6$ ② $17 + \square = 20$ ③ $\square - 25 = 8$

④ $\square - 96 = 108$ ⑤ $\square \times 3 = 6$ ⑥ $9 \times \square = 54$

⑦ $\square \div 8 = 3$ ⑧ $\square \div 4 = 6$

Supplementary Problems
•••••••• p.153

Utilization of Data

Let's interpret tables and graphs.

Want to discuss

Activity

1 The favorite menus for school lunch from the 3rd grade students in Class 1 and Class 2 were investigated and summarized in the following tables and graphs. Let's answer the following problems.

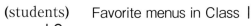

Favorite menus in Class 1

Menu	Number of students
Fried chicken	8
Hamburg steak	7
Curry and rice	5
Omelette with rice	3
Spaghetti	2
Total	25

Favorite menus in Class 2

Menu	Number of students
Fried chicken	6
Hamburg steak	7
Curry and rice	3
Omelette with rice	3
Spaghetti	4
Total	23

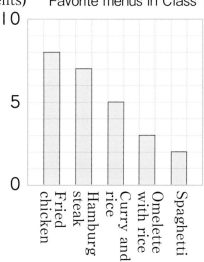

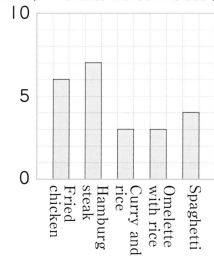

① Let's look at the tables and graphs, and discuss what you understood.

Favorite menus from 3rd grade students

(students)

Menu ＼ Class	Class 1	Class 2	Total
Fried chicken	8	6	14
Hamburg steak	7	7	14
Curry and rice	5	3	8
Omelette with rice	3	3	6
Spaghetti	2	4	6
Total	25	23	48

② The results of the investigation were summarized in one table as shown on the right. Hiroto and Nanami represented this table by using a graph as shown below. Let's compare their graphs.

Hiroto's graph

I lined up the bars.

Nanami's graph

I placed one bar on top of the other.

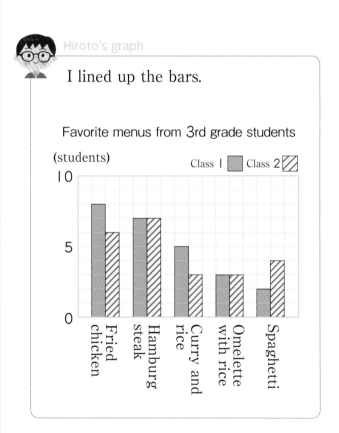

Favorite menus from 3rd grade students

Favorite menus from 3rd grade students

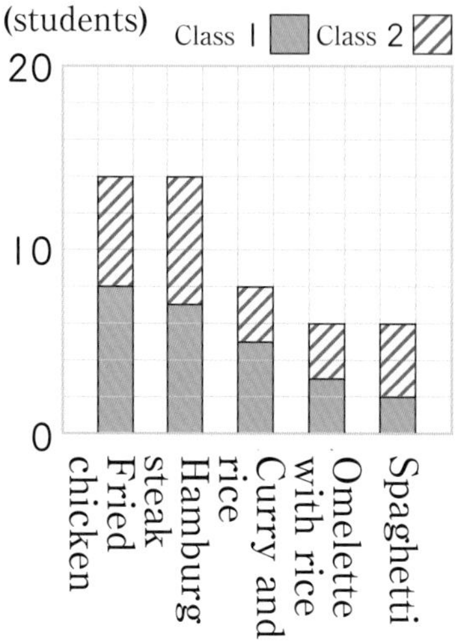

It is easier to compare the differences between Class 1 and Class 2 from Hiroto's graph.

Yui

It is easier to understand which menu is more popular from Nanami's graph.

Daiki

The graph drawn by Nanami is called a "cumulative bar graph."

Way to see and think

Both tried to find a good way to represent bar graphs for people to understand better.

2 Yui investigated about the number of books that 3rd grade students borrowed in one week at the school library. The following graph shows the number and type of books borrowed by students in each class. Let's answer the following problems.

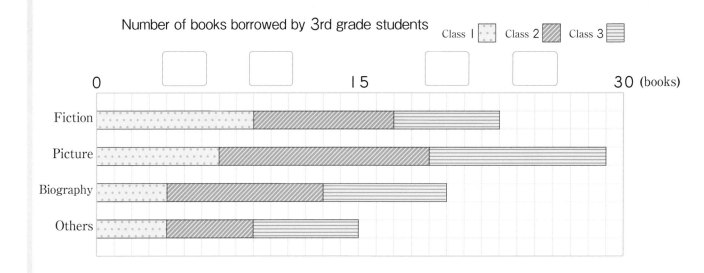

Number of books borrowed by 3rd grade students

Class 1 ☐ Class 2 ▨ Class 3 ☰

① Let's fill in each ☐ with the appropriate numbers.

② How many fiction books were borrowed by the 3rd grade students?

③ Which class borrowed the largest number of picture books? Also, how many picture books did this class borrow?

When you draw a horizontal cumulative bar graph, you should extend the bars to the right.

④ Let's organize the number of books in a table based on the bar graph from the previous page.

Number of books borrowed by 3rd grade students (books)

Type \ Class	Class 1	Class 2	Class 3	Total
Fiction				
Picture				
Biography				
Others				
Total				

1 Yui and Hiroto are looking at a graph that summarizes the number of books borrowed by 4th grade students in one week. Let's compare this graph with the 3rd grade students' graph in 2 and discuss what you noticed.

Way to see and think

It's easy to compare when everything is summarized in one table or graph.

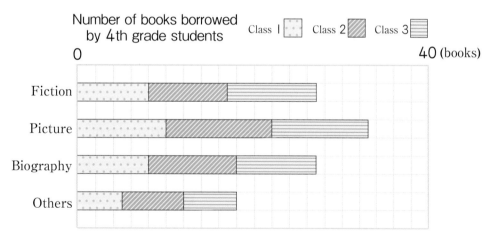

It seems that 3rd grade students read more books than 4th grade students.

Is that true? There are some differences between these two graphs.

20 Japanese Abacus
Let's think about how to represent numbers and calculate using an abacus.

1 How to represent numbers

Want to discuss

1 The abacus is an instrument used for calculations and was created a long time ago. Let's discuss how to represent numbers on it.

Ono City, Hyogo Prefecture

frame center bar rods   unit points   five-unit counters

one-unit counter

Can we represent numbers with it?

Daiki

The numerals are not written on it.

Yui

When representing numbers by using an abacus, the ones place is set on a unit point.

From right to left, the positions of the beads mean tens place, hundreds place, thousands place, etc. Also, to the right of the ones place is the tenths place.

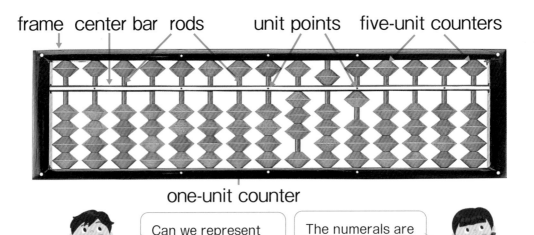

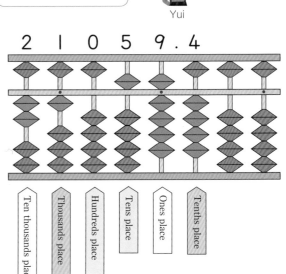

Ten thousands place | Thousands place | Hundreds place | Tens place | Ones place | Tenths place

I one-unit counter represents I and
I five-unit counter represents 5.

128

Let's read the following numbers. Also, let's fill in each ☐ with the appropriate numbers.

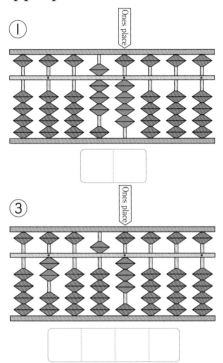

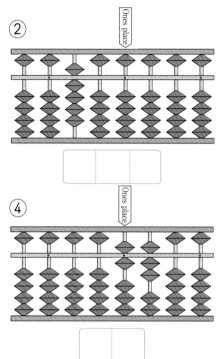

2 Let's set and clear numbers on the abacus.

	2	5	7
How to set numbers			
How to clear numbers			

Use your forefinger and thumb for setting numbers.

Only use your forefinger for clearing numbers.

3 Let's set and clear the following numbers on the abacus.

① 14　② 635　③ 500　④ 70928　⑤ 90.8

Let's learn how to add and subtract numbers by using an abacus.

① 6 + 2

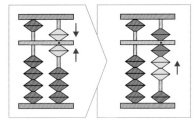

Set 6.   Add 2.

② 8 − 3

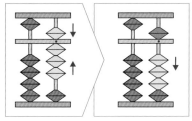

Set 8.   Subtract 3.

③ 2 + 5   ④ 1 + 7   ⑤ 5 + 4   ⑥ 6 + 3

⑦ 4 − 1   ⑧ 7 − 6   ⑨ 8 − 5   ⑩ 9 − 4

Want to confirm

 Let's calculate the following.

① 3 + 4

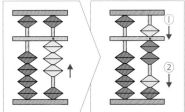

Set 3.   Add 5 and
               subtract the
               extra 1.

② 7 − 4

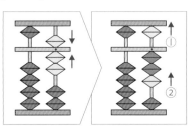

Set 7.   Subtract 5 and
               immediately
               add 1.

③ 2 + 3   ④ 3 + 3   ⑤ 4 + 2   ⑥ 4 + 3

⑦ 5 − 3   ⑧ 8 − 4   ⑨ 6 − 2   ⑩ 7 − 3

2 Let's calculate the following.

① 4 + 8　　　　　　　　　② 12 − 8

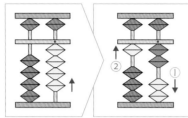

Set 4.　　Subtract 2
　　　　　and add 10.

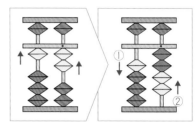

Set 12.　　Subtract 10 and
　　　　　immediately add 2.

③ 7 + 9　　④ 8 + 5　　⑤ 6 + 6　　⑥ 9 + 8

⑦ 14 − 6　　⑧ 11 − 5　　⑨ 13 − 7　　⑩ 17 − 9

Want to try

3 Let's calculate the following.

① 20 thousand + 50 thousand　　　② 1.2 − 0.4

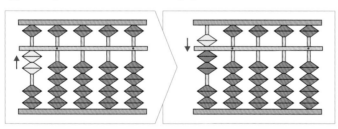

Set 20 thousand.　　Add 50 thousand.

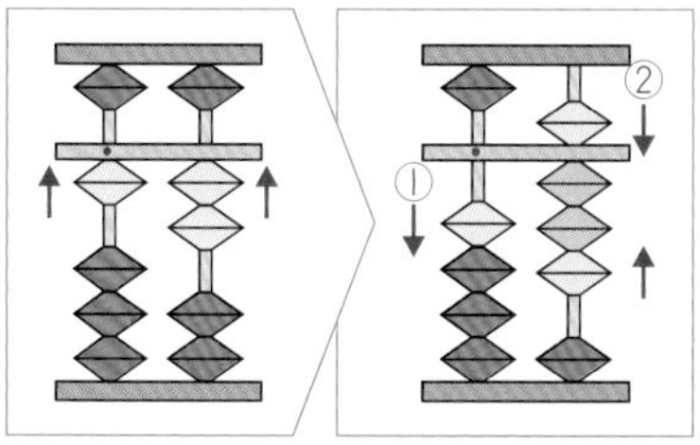

Set 1.2　　Subtract 1 and
　　　　　immediately
　　　　　add 0.6.

③ 20 thousand + 20 thousand

④ 20 thousand + 40 thousand

⑤ 40 thousand + 40 thousand

⑥ 30 thousand − 20 thousand

⑦ 50 thousand − 20 thousand

⑧ 60 thousand − 30 thousand

⑨ 0.7 + 0.2　　⑩ 0.1 + 0.5　　⑪ 1.6 + 0.6

⑫ 0.4 − 0.3　　⑬ 0.8 − 0.6　　⑭ 4.2 − 0.5

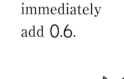

Way to see and think

All the calculations can be done in the same way, just be carefull with the place values.

131

21 3rd Grade Summary
Let's review learned mathematics.

Large Numbers, Decimal Numbers, and Fractions

1 Let's fill in each □ with the appropriate numbers.

① The numeral for the millions place in the number 58000000 is □. Also, this number gathers □ sets of ten thousand.

② 10 times of 4300 is □, and divided by 10 is □.

③ 6.7 is the sum of 6 and □. Also, this number gathers □ sets of 0.1.

④ 4 sets of $\frac{1}{7}$ is □.

Decimal Numbers and Fractions

2 Let's represent the following numbers on the number line using an ↑.

0.2 $\frac{3}{10}$ $\frac{8}{10}$ 1.6 2.1 3

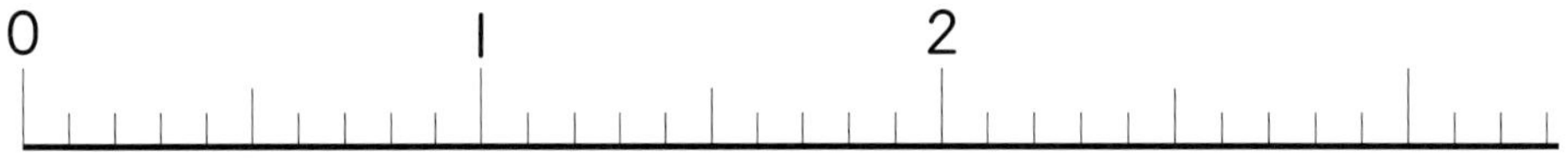

Large Numbers, Decimal Numbers, and Fractions

3 Let's fill in each □ with the appropriate equality or inequality sign.

① 32419 □ 31997 ② 0.8 □ 1.2

③ $\frac{2}{7}$ □ $\frac{6}{7}$ ④ 0.7 □ $\frac{7}{10}$

Let's calculate the following.

① 7584 + 6439   ② 671 + 4859   ③ 2901 + 99

④ 8204 − 3427   ⑤ 9002 − 7329   ⑥ 10000 − 6997

⑦ 92 × 4   ⑧ 64 × 8   ⑨ 504 × 6

⑩ 275 × 8   ⑪ 67 × 48   ⑫ 27 × 49

⑬ 30 × 70   ⑭ 870 × 32   ⑮ 508 × 50

⑯ 24 ÷ 3   ⑰ 56 ÷ 8   ⑱ 42 ÷ 7

⑲ 44 ÷ 6   ⑳ 31 ÷ 7   ㉑ 52 ÷ 9

Multiplication algorithm using squares for 56 × 82

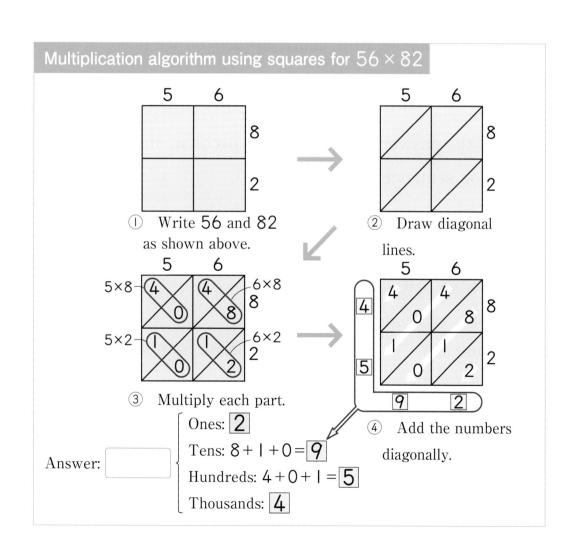

① Write 56 and 82 as shown above.

② Draw diagonal lines.

③ Multiply each part.

④ Add the numbers diagonally.

Answer:

Ones: 2
Tens: 8 + 1 + 0 = 9
Hundreds: 4 + 0 + 1 = 5
Thousands: 4

Addition and Subtraction

5 There are 5022 boys and 4828 girls in the elementary schools at the city where Rie lives.

① How many elementary school students are there altogether?

② Which is more, boys or girls, and by how many?

Multiplication with 2-digit Numbers

6 You will distribute 15 sheets of origami paper to each person. When there are 24 people, how many sheets of paper are needed altogether?

Division with Remainders

7 At the amusement park, we will ride the go-carts. Each cart has a capacity for 4 people. How many go-carts do we need for all 26 people to ride the go-carts at once?

Math Sentence using the □

8 Let's find the answer of the following problem after writing a multiplication sentence using the □.

The same number of apples were packed into 8 boxes. How many apples were packed into each box when the total number of apples was 64 apples?

Length, Time and Duration, and Weight

9 Let's fill in each ☐ with the appropriate numbers.

① 1 km = ☐ m

② 2450 m = ☐ km ☐ m

③ 1 min = ☐ sec

④ 148 sec = ☐ min ☐ sec

⑤ 9.7 kg = ☐ kg ☐ g

⑤ 3040 g = ☐ kg ☐ g

⑦ 1 kg = ☐ g

⑧ 1 L = ☐ mL

⑨ 6000 m = ☐ km

⑩ 100 mm = ☐ m

10 Let's find the following duration or time.

① The duration from 7:40 a.m. to 11:00 a.m.

② The time 1 hour 30 minutes after 10:20 a.m.

③ The children played for 1 hour 10 minutes in the morning and 50 minutes in the afternoon at the park. Let's find how long they played and the difference between the two durations.

Weight

11 The oranges were divided in two groups to be measured. How many kg and g is the total weight?

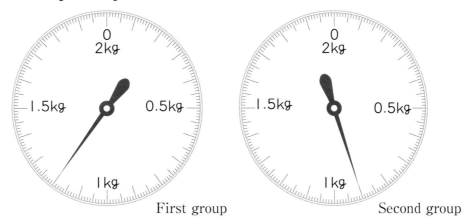

First group   Second group

Measuring the weight of an elephant

Long time ago, how did they measure the weight of an elephant?

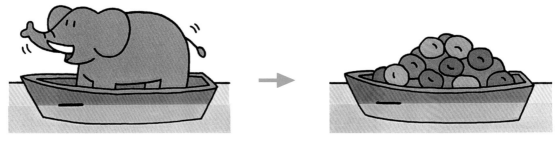

First, they placed an elephant in a boat and drew a line to show how the boat went down in the water.

Next, they placed stones in the boat until it went down to the line. Then, they measured the weight of each stone to find the total weight.

12 What are the names of the following shapes?

① A round shape drawn at the same distance from one point.

② A shape that looks like a ball and can be viewed as a circle from any direction.

③ A triangle with three equal sides.

④ A triangle with two equal sides.

13 Let's draw the following triangles. Also, what kind of triangles are these?

① A triangle whose sides have a length of 8 cm, 5 cm, and 8 cm.

② A triangle whose sides have a length of 9 cm.

14 Let's draw two circles with a radius of 4 cm and the centers at point A and B as shown on the right.

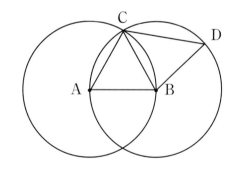

① What kind of triangle is triangle ABC? Also, let's write the reasons.

② How many cm is the length of one side of triangle ABC?

③ What kind of triangle is triangle CBD? Also, let's write the reasons.

How ancient Egyptians created right angles

A rope is divided into 12 equal parts by marking each point with a knot. A right angle can be made by the diagram shown on the right. Let's confirm if it's a right angle.

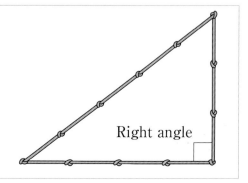

Right angle

15 The following table summarizes the grade of the students with an injury or illness who visited the school nurse during 5 days from March 1st to March 5th.

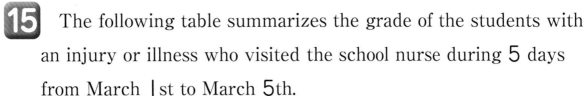

Grade of the students who visited the school nurse

March 1st	March 2nd	March 3rd	March 4th	March 5th
6th Grade	2nd Grade	2nd Grade	6th Grade	1st Grade
3rd Grade	3rd Grade	6th Grade	6th Grade	1st Grade
1st Grade	1st Grade	3rd Grade	3rd Grade	4th Grade
2nd Grade	3rd Grade	4th Grade	5th Grade	2nd Grade
3rd Grade	6th Grade	3rd Grade		3rd Grade
	4th Grade			

① Let's organize the number of students by grade who visited the school nurse into the following table.

Number of students who visited the school nurse by grade   (students)

Grade		1st	2nd	3rd	4th	5th	6th
Number of students	正						
	Numeral						

② Let's draw a bar graph using the table from ①.

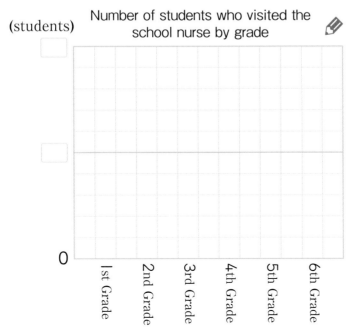

(students)

Number of students who visited the school nurse by grade

Computational thinking

01307

Let's teach Robo how to find the object with a different weight.

① There are 8 balls with the same size. Only one ball has a different weight. The only tool we have is a balance that balances when two trays have the same weight. Let's teach Robo how to find the ball with a different weight correctly.

How to find the ball with a different weight

Comparing B C D and F G H

✓ : Balancing
× : Not balancing

✓ → A or E is different

× → Comparing C D and G H

A or E is different → Comparing A and B

Comparing C D and G H ✓ → B or F is different

Comparing A and B:
✓ → E is different
× → A is different

B or F is different → Comparing B and C

Comparing B and C:
✓ → F is different
× → B is different

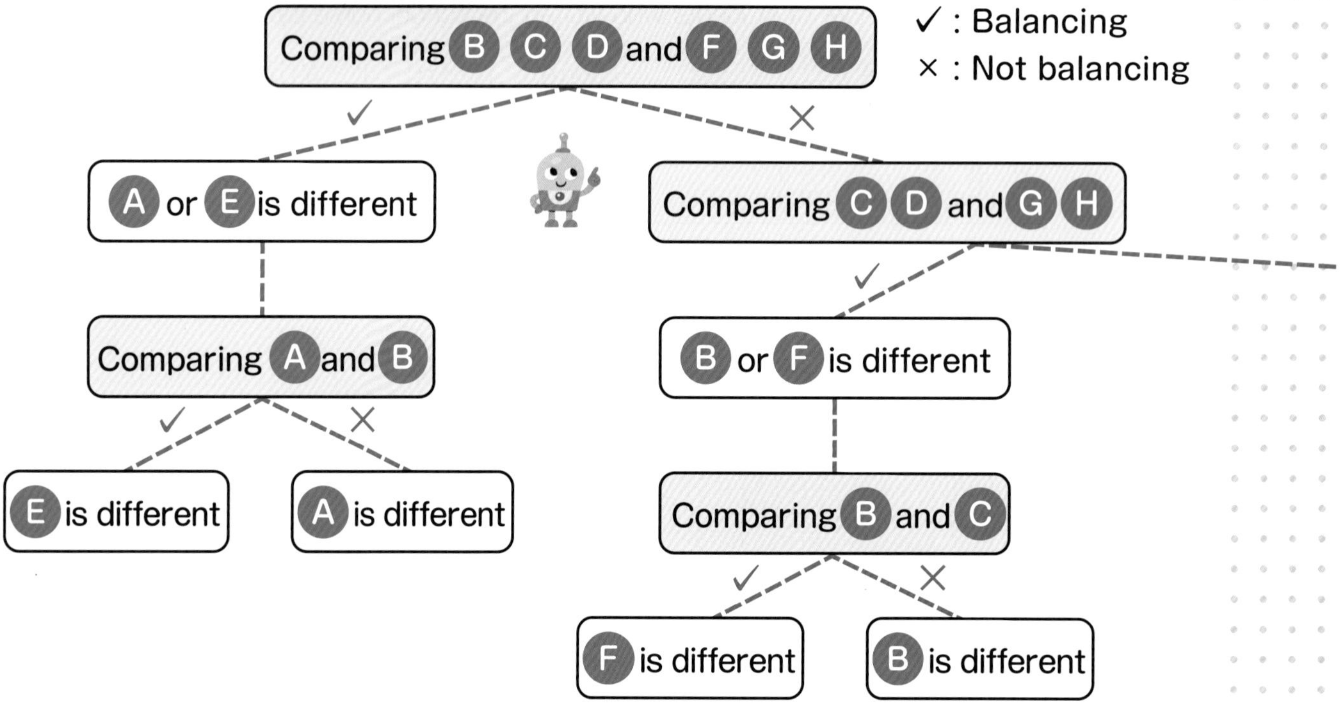

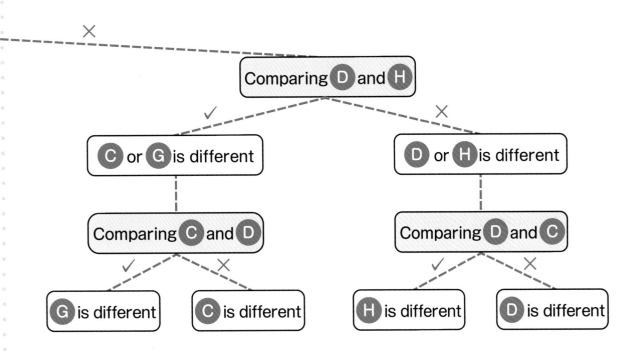

Comparing D and H

✓ C or G is different

✗ D or H is different

Comparing C and D

✓ G is different

✗ C is different

Comparing D and C

✓ H is different

✗ D is different

Utilize math for our life

Let's make Local Expert Quizzes using mathematics.

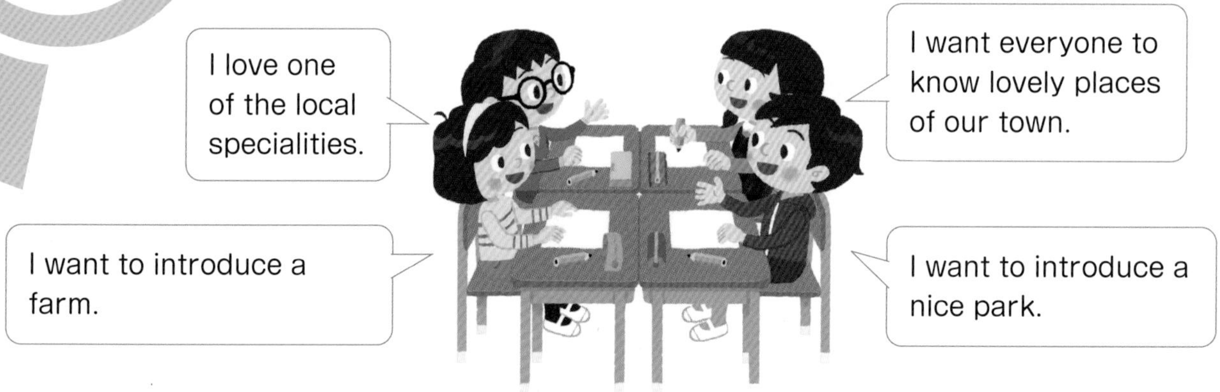

I love one of the local specialities.
I want everyone to know lovely places of our town.
I want to introduce a farm.
I want to introduce a nice park.

Let's investigate about our town and make quizzes.

1. Local Park Course

Quiz 1: Big Pine Tree

The map of Park X is shown on the right. We walk into the park from the entrance and go to the exit visiting a big tree, a bench, a small island, and a pavilion on the way. Is it possible to visit all these places without passing through the same route?

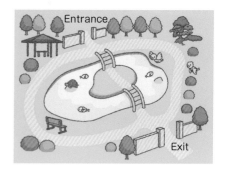

Entrance
Exit

? Let's make quizzes about famous places in our town such as tourist attractions.

2. Local Shop Course

Quiz 2: Supermarket

Which amount of water is heavier, the amount of water inside one bottle of 1L or the amount of water inside two bottles of 500mL?

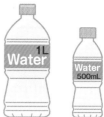

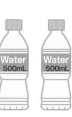

1L
Water
Water
500mL
Water
500mL

? Let's make quizzes about shops that you visit frequently.

3. Local Food Place Course

Quiz 3: Factory

One of our local food places is the cake factory. They have two types of boxes. The first one comes with 12 cakes and the second one comes with 8 cakes. The factory is capable of making 300 boxes for each type in 1 hour. How many cakes can they make altogether in 1 hour?

? Let's make quizzes based on local factories or shops.

4. Local Industry Course

Quiz 4: Farm

Watermelons, melons, and strawberries are well cultivated in our town. We investigated about other prefectures that cultivate these fruits in Japan and found the best five for the year 2015. Which of the following graphs represent the watermelon, melon, and strawberry?

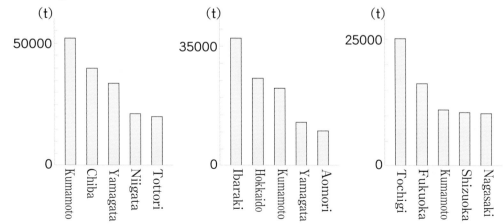

? Let's make quizzes based on research on crops or other things that are made in our town.

Utilize math for our life

Let's make Local Expert Quizzes using mathematics.

1. Toward learning competency

	😊 Strongly agree	🙂 Agree	☹ Don't agree
① It was fun making quizzes.			
② The learning contents were helpful.			
③ I made quizzes by my own initiative.			

2. Thinking, deciding, and representing competency

	😊 Definitely did	🙂 So so	☹ I didn't
① I was able to develop local quizzes using mathematics.			
② I was able to confirm whether I can solve the quizzes.			
③ I was able to represent the quizzes using sentences, pictures, and diagrams.			

3. What I know and can do

	😊 Definitely did	🙂 So so	☹ I didn't
① I was able to make better quizzes.			
② I was able to solve the quizzes.			

4. Encouragement for myself

	😊 Strongly agree
① I think that I'm doing my best.	

Give yourself a compliment since you have worked so hard.

Let's try to work out what you were not able to accomplish and keep doing your best on what you were able to fulfill. Try to make another quiz by changing the theme.

Supplementary Problems

⑩ Large Numbers

pp. 4~19

1 Let's write the following numbers in numerals.

① sixty-four thousand eight hundred fifty

② ninety thousand thirty-one

③ twenty thousand seven hundred four

2 Let's fill in each ☐ with the appropriate numbers.

① 53000 is the sum of ☐ sets of ten thousand and ☐ sets of one thousand.

② 90840 is the sum of ☐ sets of ten thousand, ☐ sets of one hundred, and ☐ sets of ten.

③ ☐ is the sum of 2 sets of ten thousand, 7 sets of one thousand, and 1 set of ten.

④ ☐ is the sum of 6 sets of ten thousand and 9 sets of one hundred.

3 Let's read the following numbers.

① 462900 ② 7058280

③ 13090600

4 Let's fill in each ☐ with the appropriate numbers.

① 470000 is the sum of ☐ sets of one hundred thousand and ☐ sets of ten thousand.

② 28050000 is the sum of ☐ sets of ten million, ☐ sets of one million, and ☐ sets of ten thousand.

③ ☐ is the sum of 6 sets of one hundred thousand and 7 sets of ten thousand.

④ ☐ is the sum of 9 sets of one million, 1 set of ten thousand, and 2 sets of one thousand.

5 The following numbers gather how many sets of 1000? Also, how many sets of 100?

① 36000 ② 490000

6 Let's find the following numbers.

① The number that is 10 times of 26
② The number that is 10 times of 875
③ The number that is 100 times of 690
④ The number that is 100 times of 32
⑤ The number that is 1000 times of 506

7 Let's divide the following numbers by 10.

① 80 ② 190 ③ 600 ④ 550

8 Let's write the numbers represented by ① ~ ④.

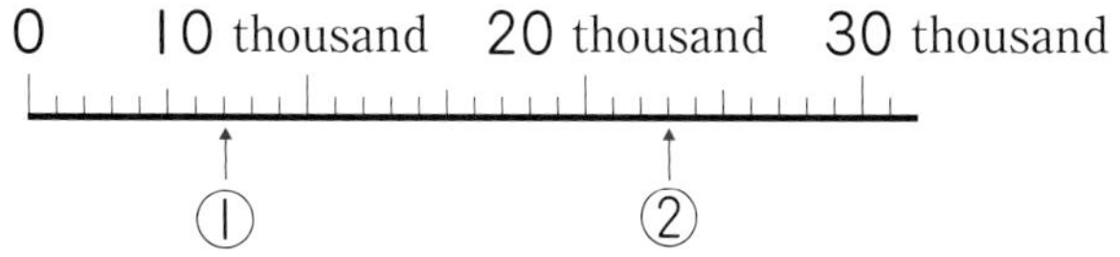

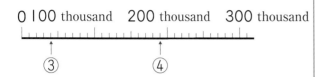

0　　10 thousand　20 thousand　30 thousand

①　　②

0 100 thousand　200 thousand　300 thousand

③　　④

9 Let's fill in each ▢ with the appropriate numbers.

① ┤99600├ ┤99800├
┤ ├ ┤100200├

② ┤5 million 900 thousand├ ┤ ├
┤6 million├┤6 million 50 thousand├

③ ┤ ├ ┤6 million├
┤8 million├┤10 million├

④ ┤8 million 600 thousand├
┤8 million 800 thousand├ ┤ ├
┤9 million 200 thousand├

10 Let's fill in each ▢ with the appropriate inequality sign.

① 60842 ▢ 520346

② 81620 ▢ 79183

③ 49700 ▢ 49080

④ 53482 ▢ 54382

11 Let's calculate the following.

① 330000 + 260000

② 640000 − 190000

③ 1 million 780 thousand + 1 million 150 thousand

④ 8 million 920 thousand − 2 million 230 thousand

⑤ 30 million + 70 million

⑥ 45 million − 14 million

⑪ Circles and Spheres

pp. 20~36

1 Let's write appropriate words for Ⓐ, Ⓑ, and Ⓒ in the circle shown below.

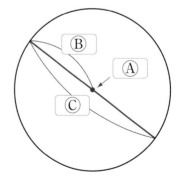

2 The length of the diameter is how many times the length of the radius?

3 A circle is drawn to fit exactly inside a square with a side of 10 cm. How many cm is the radius?

④ Let's draw a circle with a radius of 4 cm.

⑤ Let's use a compass to compare the length of the following straight lines and arrange them from longest to shortest.

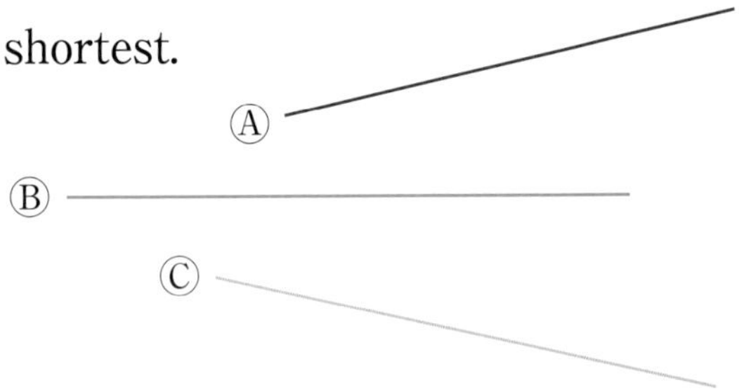

Ⓐ

Ⓑ ——————————

Ⓒ

⑥ Each of the two circles shown below has a diameter of 10 cm. How many cm is the length of straight line AB?

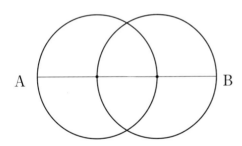

A   B

⑦ The three circles shown below have the same size. How many cm is the diameter of each circle when the length of straight line AB is 28 cm?

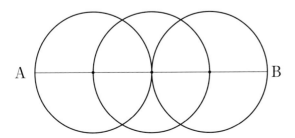

A   B

⑫ Time and Duration (2)
pp. 37~40

① Let's fill in each ▢ with the appropriate numbers.
① 1 min 25 sec = ▢ sec
② 1 min 52 sec = ▢ sec
③ 97 sec = ▢ min ▢ sec
④ 108 sec = ▢ min ▢ sec

⑬ Decimal Numbers
pp. 41~53

① How many dL are there in the following amounts of water?

①   ②

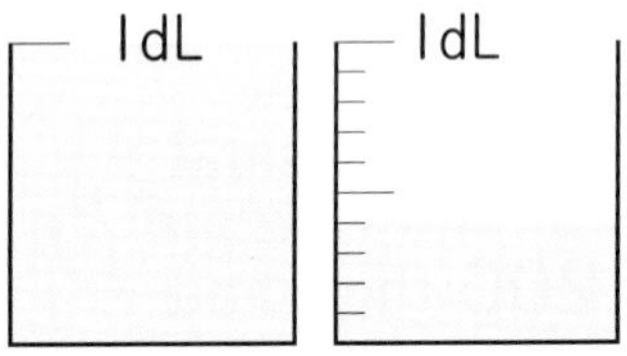

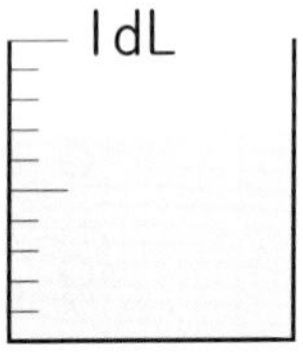

⌐ 1dL   ⌐ 1dL   ⌐ 1dL

② Let's fill in each ▢ with the appropriate numbers.
① 8 sets of 0.1 dL is ▢ dL.
② ▢ sets of 0.1 dL is 1.2 dL.
③ 34 sets of 0.1 dL is ▢ dL.
④ 4 sets of 1 dL and 5 sets of 0.1 dL make ▢ dL.

3 How many cm are the following lengths? Let's represent with decimal numbers.
① 4 mm ② 1 cm 7 mm
③ 3 cm 8 mm ④ 5 cm 5 mm

4 How many m are the following lengths? Let's represent with decimal numbers.
① 20 cm ② 90 cm
③ 1 m 80 cm ④ 2 m 60 cm

5 Let's write the numbers that are pointed on the number line with an ↑.

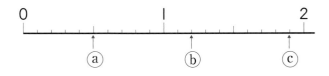

6 Let's fill in each ☐ with the appropriate numbers.
① 2.5 is ☐ sets of 0.1.
② 0.7 is ☐ sets of 0.1.
③ 18 sets of 0.1 is ☐.

7 Let's fill in each ☐ with the appropriate inequality sign.
① 6 ☐ 6.2 ② 5.4 ☐ 5
③ 6.8 ☐ 4.5 ④ 8.3 ☐ 7.9
⑤ 2.3 ☐ 3.1 ⑥ 0.9 ☐ 1.1

8 Let's calculate the following.
① 0.4 + 0.2 ② 0.5 + 0.4
③ 0.8 + 0.8 ④ 0.5 + 0.7

9 Let's calculate the following in vertical form.
① 3.2 + 2.6 ② 1.4 + 3.5
③ 1.8 + 4.4 ④ 4.3 + 1.7
⑤ 3.9 + 0.5 ⑥ 0.6 + 1.4
⑦ 3 + 2.4 ⑧ 1.5 + 7

10 Let's calculate the following.
① 0.8 − 0.4 ② 0.7 − 0.2
③ 1.7 − 0.9 ④ 1.4 − 0.7

11 Let's calculate the following in vertical form.
① 3.8 − 1.4 ② 6.7 − 3.4
③ 3.4 − 1.6 ④ 4.5 − 2.8
⑤ 5.4 − 2.5 ⑥ 5 − 2.8
⑦ 2 − 1.2 ⑧ 3 − 1.6

12 There are 2 cups that contain 2.4 dL of juice and 1.8 dL of juice respectively. How many dL are there altogether?

13 There was a 2.3 m ribbon and 1.5 m of it were used. How many m are left?

⑭ Triangles and Angles

pp.54～69

① Which of the following triangles are isosceles triangles?

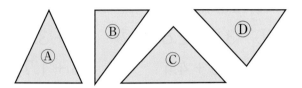

② Which of the following triangles are equilateral triangles?

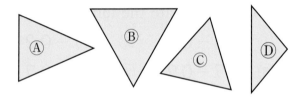

③ Let's draw the following triangles.

① An isosceles triangle whose sides have a length of 4 cm, 5 cm, and 5 cm.

② An isosceles triangle whose sides have a length of 6 cm, 6 cm, and 7 cm.

④ Let's draw the following triangles.

① An equilateral triangle whose sides have a length of 2 cm.

② An equilateral triangle whose sides have a length of 6 cm.

⑤ Let's answer the following about the circle with a radius of 4 cm shown on the right.

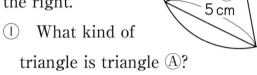

① What kind of triangle is triangle Ⓐ?

② What kind of triangle is triangle Ⓑ?

⑥ There are angles Ⓐ ～ Ⓓ as shown below. Let's arrange them in descending order.

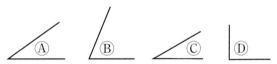

⑦ In the following triangles, which angles have the same size?

① Isosceles triangle

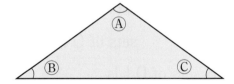

② Equilateral triangle

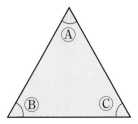

148

⑮ Multiplication with 2-digit Numbers

pp.72~84

1 Let's fill in each ☐ with the appropriate numbers.

① $3 \times 70 = 3 \times 7 \times \boxed{}$

$= 21 \times \boxed{}$

$= \boxed{}$

② $20 \times 90 = 2 \times 9 \times \boxed{} \times \boxed{}$

$= 2 \times 9 \times \boxed{}$

$= 18 \times \boxed{}$

$= \boxed{}$

2 Let's calculate the following.

① 4×70　　② 9×40

③ 6×50　　④ 60×20

⑤ 90×30　　⑥ 50×40

3 Let's fill in each ☐ with the appropriate numbers regarding how to calculate 32×12.

32×12 ⟨ $32 \times 2 = \boxed{}$
$32 \times \boxed{} = \boxed{}$

Total $\boxed{}$

4 Let's calculate the following in vertical form.

① 22×22　　② 36×11

③ 12×23　　④ 11×72

5 Let's fill in each ☐ with the appropriate numbers.

$$\begin{array}{r} 3\ 8 \\ \times\ 2\ 6 \\ \hline \boxed{\ }\boxed{\ }\boxed{\ } \end{array} \rightarrow \begin{array}{r} 3\ 8 \\ \times\ 2\ 6 \\ \hline 2\ 2\ 8 \\ \boxed{\ }\boxed{\ } \end{array} \rightarrow \begin{array}{r} 3\ 8 \\ \times\ 2\ 6 \\ \hline 2\ 2\ 8 \\ 7\ 6\ \\ \hline \boxed{\ }\boxed{\ }\boxed{\ } \end{array}$$

6 Let's calculate the following in vertical form.

① 15×12　　② 24×13

③ 26×31　　④ 27×24

⑤ 19×38　　⑥ 46×17

7 You will distribute 15 sheets of origami paper to each person. When there are 36 people, how many sheets of paper are needed altogether?

8 Let's calculate the following in vertical form.

① 57×87　　② 74×86

③ 44×48　　④ 46×97

⑤ 76×83　　⑥ 25×79

9 Let's calculate the following in vertical form.

① 23×40　　② 35×60

③ 30×13　　④ 90×14

10 You bought 20 sets of snacks that cost 95 yen each. How many yen is the total cost?

⑪ Let's fill in each ☐ with the appropriate numbers regarding how to calculate 132×32.

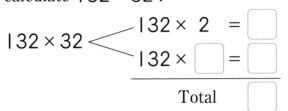

$$132 \times 2 = \boxed{}$$
$$132 \times \boxed{}\ = \boxed{}$$

$$132 \times 32 \left\{ \begin{array}{l} \end{array} \right.$$

Total $\boxed{}$

⑫ Let's calculate the following in vertical form.
① 231×32 ② 322×12
③ 312×13 ④ 222×32
⑤ 132×23 ⑥ 212×43

⑬ Let's calculate the following in vertical form.
① 418×68 ② 894×41
③ 337×85 ④ 684×58
⑤ 615×26 ⑥ 940×25

⑭ Let's calculate the following in vertical form.
① 409×40 ② 703×80
③ 604×30 ④ 802×26
⑤ 600×43 ⑥ 900×70

⑮ Let's find the mistakes in the following calculations and calculate correctly.

①
```
    6 2
  × 4 3
  ─────
  1 8 6
  2 4 8
  ─────
  4 3 4
```

②
```
    3 7 4
  ×    5 0
  ───────
  1 8 7 0
```

⑯ Fractions
pp.85~98

❶ How many m is the length of the following colored parts?

①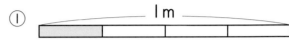

②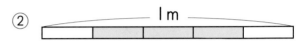

③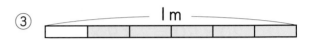

❷ How many L are the following amounts of water?

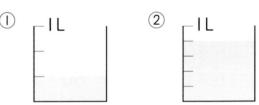

❸ Let's fill in each ☐ with the appropriate numbers.

① $\dfrac{3}{5}$ m is $\boxed{}$ sets of $\dfrac{1}{5}$ m.

② 1 L is $\boxed{}$ sets of $\dfrac{1}{3}$ L.

③ $\dfrac{\boxed{}}{5} = 1$

④ Comparing $\dfrac{7}{8}$ m and 1 m, $\boxed{}$ m is longer.

④ Let's fill in each ☐ with the appropriate inequality sign.

① $\dfrac{2}{5}$ ☐ $\dfrac{1}{5}$ ② $\dfrac{4}{6}$ ☐ $\dfrac{5}{6}$

③ $\dfrac{5}{7}$ ☐ $\dfrac{4}{7}$ ④ $\dfrac{1}{2}$ ☐ 1

⑤ Let's represent numbers ⓐ and ⓑ as fractions and numbers ⓒ and ⓓ as decimal numbers.

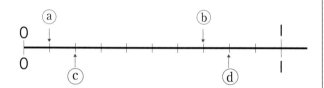

⑥ Let's represent the following fractions as decimal numbers.

① $\dfrac{4}{10}$ ② $\dfrac{7}{10}$

⑦ Let's represent the following decimal numbers as fractions.

① 0.3 ② 0.9

⑧ Let's fill in each ☐ with the appropriate equality or inequality sign.

① $\dfrac{3}{10}$ ☐ 0.4 ② 0.8 ☐ $\dfrac{6}{10}$

③ $\dfrac{9}{10}$ ☐ 0.9 ④ 0.5 ☐ $\dfrac{3}{10}$

⑨ Let's calculate the following.

① $\dfrac{1}{3} + \dfrac{1}{3}$ ② $\dfrac{2}{9} + \dfrac{5}{9}$

③ $\dfrac{2}{5} + \dfrac{2}{5}$ ④ $\dfrac{3}{8} + \dfrac{2}{8}$

⑤ $\dfrac{1}{2} + \dfrac{1}{2}$ ⑥ $\dfrac{5}{7} + \dfrac{2}{7}$

⑩ Marina used $\dfrac{4}{8}$ m of a ribbon yesterday and $\dfrac{3}{8}$ m today. How many m of ribbon did she use altogether?

⑪ Let's calculate the following.

① $\dfrac{3}{4} - \dfrac{2}{4}$ ② $\dfrac{5}{7} - \dfrac{2}{7}$

③ $\dfrac{4}{5} - \dfrac{3}{5}$ ④ $\dfrac{3}{5} - \dfrac{2}{5}$

⑤ $\dfrac{7}{8} - \dfrac{3}{8}$ ⑥ $1 - \dfrac{2}{3}$

⑫ There is 1 L of orange juice. When we drink $\dfrac{4}{9}$ L of juice, how many L of juice are left?

⑰ Weight

pp.99~113

1 Let's answer the following problems.

① When 2 kg of sand and 3 kg of sand are added, how many kg are there altogether?

② When 1 L of water and 3 L of water are added, how many L are there altogether? Also, how many kg is the total weight?

③ How many g is 1 kg?

④ How many kg is 1 t?

2 How many kg and g are the weights shown below? Also, how many g?

①

②

3 How many kg and g are the weights shown below?

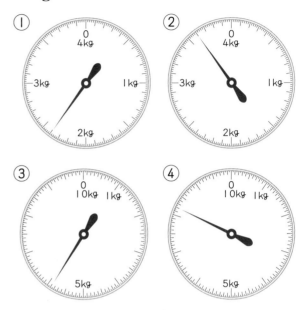

4 Let's fill in each ☐ with the appropriate numbers.

① 2 kg 360 g = ☐ g

② 3 kg 80 g = ☐ g

③ 4.6 kg = ☐ kg ☐ g

④ 1500 g = ☐ kg

5 After placing oranges in a basket that weighs 300 g, the total weight became 1 kg 150 g.

How many g is the weight of the oranges?

⑱ Math Sentences Using the □ pp.116~123

1 There was an initial number of sheets of paper. After receiving 15 sheets of paper, the total number of sheets of paper became 50 sheets of paper. Let's answer the following problems.

① We are representing this as a math sentence using words. Let's fill in the □ with the appropriate words.

$$\boxed{\text{Initial number of sheets of paper}} + \boxed{}$$
$$= \boxed{\text{Total number of sheets of paper}}$$

② Let's represent this as a math sentence for finding the total number of sheets of paper by using □ sheets as the initial number of sheets of paper.

③ Let's find the initial number of sheets of paper by using the math sentence from ②.

2 Let's find the number that applies to each □.

① $\square + 16 = 40$ ② $\square + 47 = 92$
③ $38 + \square = 63$ ④ $53 + \square = 120$

3 Momoka is reading a book. Until now, she has read 78 pages of the book, but 42 pages are remaining.

Let's answer the following problems.

① We are representing this as a math sentence using words. Let's fill in the □ with the appropriate words.

$$\boxed{\text{Number of pages of the book}} - \boxed{}$$
$$= \boxed{\text{Remaining number of pages}}$$

② Let's represent this as a math sentence for finding the remaining number of pages by using □ pages as the number of pages of the book.

③ Let's find the number of pages of the book by using the math sentence from ②.

4 Let's find the number that applies to each □.

① $\square - 46 = 32$ ② $\square - 29 = 66$
③ $\square - 78 = 123$ ④ $\square - 160 = 90$

5 4 notebooks with the same price were bought for 360 yen. Let's answer the following problems.

① Let's represent this as a math sentence for finding the total cost by using □ yen as the cost for one notebook.

② Let's find the cost for one notebook by using the math sentence from ①.

6 Let's find the number that applies to each □.

① $\square \times 6 = 30$ ② $8 \times \square = 64$
③ $\square \div 9 = 7$

Answers

10 Large Numbers

1 ① 64850 ② 90031 ③ 20704

2 ① 5, 3 ② 9, 8, 4 ③ 27010 ④ 60900

3 ① four hundred sixty-two thousand nine hundred

② seven million fifty-eight thousand two hundred eighty

③ thirteen million ninety thousand six hundred

4 ① 4, 7 ② 2, 8, 5 ③ 670000 ④ 9012000

5 ① 1000…36 sets 100…360 sets

② 1000…490 sets 100…4900 sets

6 ① 260 ② 8750 ③ 69000 ④ 3200 ⑤ 506000

7 ① 8 ② 19 ③ 60 ④ 55

8 ① 7000 ② 23000 ③ 40 thousand ④ 190 thousand

9 ① 100000 ② 5 million 950 thousand ③ 4 million ④ 9 million

10 ① < ② > ③ > ④ <

11 ① 590000 ② 450000 ③ 2 million 930 thousand

④ 6 million 690 thousand ⑤ 100 million ⑥ 31 million

11 Circles and Spheres

1 Ⓐ center Ⓑ radius Ⓒ diameter

2 2 times

3 5 cm

4 (Omitted)

5 Ⓑ, Ⓒ, Ⓐ

6 15 cm

7 14 cm

12 Time and Duration (2)

1 ① 85 ② 112 ③ 1, 37 ④ 1, 48

13 Decimal Numbers

1 ① 1.4 dL ② 0.7 dL

2 ① 0.8 ② 12 ③ 3.4 ④ 4.5

3 ① 0.4 cm ② 1.7 cm ③ 3.8 cm ④ 5.5 cm

4 ① 0.2 m ② 0.9 m ③ 1.8 m ④ 2.6 m

5 ⓐ 0.5 ⓑ 1.2 ⓒ 1.9

6 ① 25 ② 7 ③ 1.8

7 ① < ② > ③ > ④ > ⑤ < ⑥ <

8 ① 0.6 ② 0.9 ③ 1.6 ④ 1.2

9 ① 5.8 ② 4.9 ③ 6.2 ④ 6 ⑤ 4.4 ⑥ 2

⑦ 5.4 ⑧ 8.5

10 ① 0.4 ② 0.5 ③ 0.8 ④ 0.7

11 ① 2.4 ② 3.3 ③ 1.8 ④ 1.7 ⑤ 2.9 ⑥ 2.2

⑦ 0.8 ⑧ 1.4

12 2.4 + 1.8 = 4.2 4.2 dL

13 2.3 − 1.5 = 0.8 0.8 m

14 Triangles and Angles

1 Ⓐ, Ⓒ

2 Ⓑ, Ⓒ

3 (omitted)

4 (omitted)

5 ① isosceles triangle ② equilateral triangle

6 Ⓓ, Ⓑ, Ⓐ, Ⓒ

7 ① Ⓑ and Ⓒ ② Ⓐ, Ⓑ, and Ⓒ

15 Multiplication with 2-digit Numbers

1 ① 10, 10, 210 ② 10, 10, 100, 100, 1800

2 ① 280 ② 360 ③ 300 ④ 1200 ⑤ 2700

⑥ 2000

3

$$32 \times 12 \begin{cases} 32 \times 2 = \boxed{64} \\ 32 \times \boxed{10} = \boxed{320} \end{cases}$$

Total $\boxed{384}$

4 ① 484 ② 396 ③ 276 ④ 792

5 228, 76, 988

6 ① 180 ② 312 ③ 806 ④ 648 ⑤ 722

⑥ 782

7 15 × 36 = 540 540 sheets of paper

8 ① 4959 ② 6364 ③ 2112 ④ 4462

⑤ 6308 ⑥ 1975

9 ① 920 ② 2100 ③ 390 ④ 1260

10 95 × 20 = 1900 1900 yen

11

$$132 \times 32 \begin{cases} 132 \times 2 = \boxed{264} \\ 132 \times \boxed{30} = \boxed{3960} \end{cases}$$

Total $\boxed{4224}$

12 ① 7392 ② 3864 ③ 4056 ④ 7104

⑤ 3036 ⑥ 9116

13 ① 28424 ② 36654 ③ 28645 ④ 39672

⑤ 15990 ⑥ 23500

14 ① 16360 ② 56240 ③ 18120 ④ 20852

⑤ 25800 ⑥ 63000

15 ①　　　　　　　②

```
      6 2              3 7 4
    × 4 3            ×   5 0
    ─────            ───────
    1 8 6            1 8 7 0 0
  2 4 8
  ───────
  2 6 6 6
```

⑯ Fractions

1 ① $\frac{1}{4}$ m ② $\frac{3}{5}$ m ③ $\frac{5}{6}$ m

2 ① $\frac{1}{3}$ L ② $\frac{4}{5}$ L

3 ① 3 ② 3 ③ 5 ④ 1

4 ① > ② < ③ > ④ <

5 ⓐ $\frac{1}{10}$ ⓑ $\frac{7}{10}$ ⓒ 0.2 ⓓ 0.8

6 ① 0.4 ② 0.7

7 ① $\frac{3}{10}$ ② $\frac{9}{10}$

8 ① < ② > ③ = ④ >

9 ① $\frac{2}{3}$ ② $\frac{7}{9}$ ③ $\frac{4}{5}$ ④ $\frac{5}{8}$ ⑤ 1 ⑥ 1

10 $\frac{4}{8} + \frac{3}{8} = \frac{7}{8}$ $\underline{\frac{7}{8}}$ m

11 ① $\frac{1}{4}$ ② $\frac{3}{7}$ ③ $\frac{1}{5}$ ④ $\frac{1}{5}$ ⑤ $\frac{4}{8}$ ⑥ $\frac{1}{3}$

12 $1 - \frac{4}{9} = \frac{5}{9}$ $\underline{\frac{5}{9}}$ L

⑰ Weight

1 ① 5 kg ② 4 L, 4 kg ③ 1000 g ④ 1000 kg

2 ① 1 kg 300 g, 1300 g ② 1 kg 650 g, 1650 g

3 ① 2 kg 400 g ② 3 kg 600 g ③ 5 kg 900 g
④ 8 kg 200 g

4 ① 2360 ② 3080 ③ 4, 600 ④ 1.5

5 1150 − 300 = 850 $\underline{850 \text{ g}}$

⑱ Math Sentences Using the □

1 ① Number of sheets received ② □＋15 = 50 ③ 35 sheets

2 ① 24 ② 45 ③ 25 ④ 67

3 ① Number of pages read
② □− 78 = 42 ③ 120 pages

4 ① 78 ② 95 ③ 201 ④ 250

5 ① □× 4 = 360 ② 90 yen

6 ① 5 ② 8 ③ 63

Symbols and words in this book

Isosceles triangles ▼ will be used in page 66.

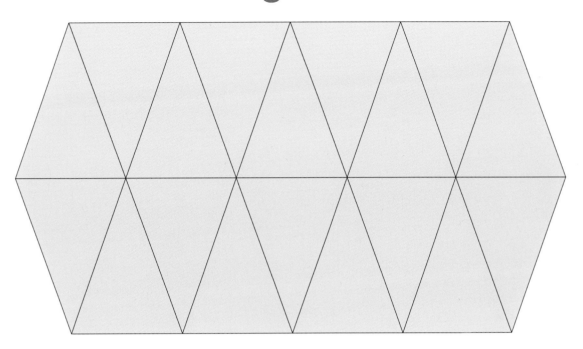

Equilateral triangles ▼ will be used in page 66.

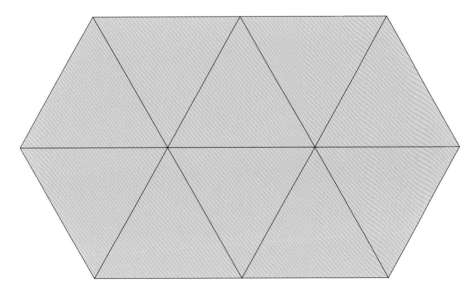

▼ The triangles on this page will be cut and used in pages 56 and 57.

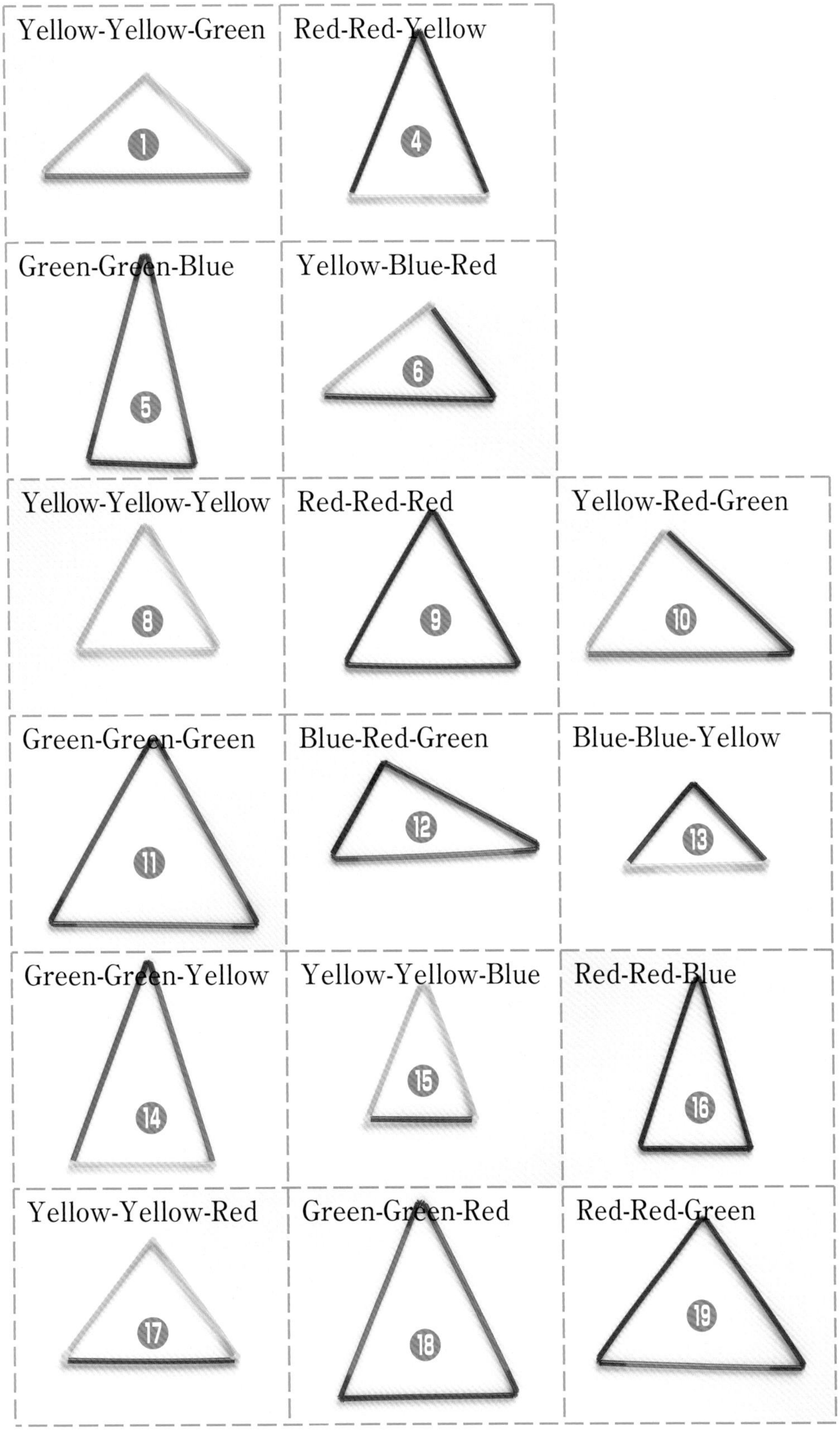

Yellow-Yellow-Green 1

Red-Red-Yellow 4

Green-Green-Blue 5

Yellow-Blue-Red 6

Yellow-Yellow-Yellow 8

Red-Red-Red 9

Yellow-Red-Green 10

Green-Green-Green 11

Blue-Red-Green 12

Blue-Blue-Yellow 13

Green-Green-Yellow 14

Yellow-Yellow-Blue 15

Red-Red-Blue 16

Yellow-Yellow-Red 17

Green-Green-Red 18

Red-Red-Green 19

Memo

Memo

Memo

Editorial for English Edition:

Study with Your Friends, Mathematics for Elementary School
3rd Grade, Vol.2, Gakko Tosho Co.,Ltd., Tokyo, Japan [2020]